AA007567

V International Scientific Conference "Survivability and Structural Material Science" (SSMS 2020)

IOP Conference Series: Materials Science and Engineering
Volume 1023

Moscow, Russia
27 - 29 October 2020

ISBN: 978-1-7138-3324-6
ISSN: 1757-8981

Printed from e-media with permission by:

Curran Associates, Inc.
57 Morehouse Lane
Red Hook, NY 12571

Some format issues inherent in the e-media version may also appear in this print version.

Institute of Physics
Dirac House, Temple Back
Bristol BS1 6BE UK

Phone: 44 1 17 929 7481
Fax: 44 1 17 920 0979

techtracking@iop.org

Additional copies of this publication are available from:

Curran Associates, Inc.
57 Morehouse Lane
Red Hook, NY 12571 USA
Phone: 845-758-0400
Fax: 845-758-2633
Email: curran@proceedings.com
Web: www.proceedings.com

TABLE OF CONTENTS

V International Scientific Conference "Survivability and Structural Material Science" (SSMS 2020), 27-29 October 2020, Moscow, Russia.

We are pleased to introduce the Proceedings of the V International Scientific Conference "Survivability and Structural Material Science" (SSMS 2020).

The Conference was held from October 27th to October 29th, 2020 in an online format. It was organized by Blagonravov Mechanical Engineering Research Institute of the Russian Academy of Sciences (IMASH RAS).

The scientific sections of the Conference covered a wide range of topics on fracture mechanics, structural materials science, survivability, engineering safety and risk analysis, structural integrity, tribology and surface coatings.

In 2020, the conference was held virtually due to restrictions related to COVID19 using the ZOOM platform. The organizers were located at the Blagonravov Mechanical Engineering Research Institute of the Russian Academy of Sciences, Moscow, Russia.

The meetings were held in the format of a speech with an oral presentation. For each participant was allotted 15 minutes including questions. Plenary sessions were held on the first day of the conference.

91 scientists from various cities of Russia, France and Germany participated in the conference in 2020. The organizing committee selected 29 articles of the participants for publication in the Proceeding of conference.

The conference passed without major problems. Organizing committee connected many participants virtually. The exhibition part of the conference was impossible due to the remote presentation.

List of Conference Committee

Ganiev R. F., Honorary Conference Chairman, academician of RAS, Scientific Director of IMASH RAN, Mechanical Engineering Research Institute of the Russian Academy of Sciences (IMASH RAN), Moscow, Russia

Glazunov V. A., Chairman of Organizing Committee, Ph.D., Director of IMASH RAN, Mechanical Engineering Research Institute of the Russian Academy of Sciences (IMASH RAN), Moscow, Russia

Matvienko Yu. G., Ph.D., IMASH RAN, Deputy of Chairman, Moscow, Russia

Makhutov N. A., Corresponding Member of RAS, Deputy of Chairman, Moscow, Russia

Romanov A. N., Ph.D., IMASH RAN, Deputy of Chairman, Moscow, Russia

Albagachiev A.Yu., IMASH RAN, Moscow, Russia

Alisin V.V., Ph.D., IMASH RAN, Moscow, Russia

Antonova O S - IMASH RAN, Moscow, Russia

Azikov N.S., Ph.D., IMASH RAN, Moscow, Russia

Barinov S M - Corresponding Member of RAS, IMET RAN, Moscow, Russia

Biryukov V.P., Ph.D., IMASH RAN, Moscow, Russia

Dumansky A.M., Ph.D., IMASH RAN, Moscow, Russia

Gadolina I.V., IMASH RAN, Moscow, Russia

Kalita V. I., Ph.D., IMET RAN, Moscow, Russia

Khrushchev M.M., IMASH RAN, Moscow, Russia

Levashov E.A., Ph.D., MISIS, Moscow, Russia

Levin I.S., Ph.D., IMASH RAN, Moscow, Russia

Lomakin E.V., Corresponding Member of RAS, MSU, Moscow, Russia

Polilov A.N., Ph.D., IMASH RAN, Moscow, Russia

Prozhega M.V., IMASH RAN, Moscow, Russia

Pugachev M.S., IMASH RAN, Moscow, Russia

Stolyarov V.V., Ph.D., IMASH RAN, Moscow, Russia

Tatous' N.A., Ph.D., IMASH RAN, Moscow, Russia

Yudkin V.F., IMASH RAN, Moscow, Russia

Peer review declaration

All papers published in this volume of *IOP Conference Series: Materials Science and Engineering* have been peer reviewed through processes administered by the Editors. Reviews were conducted by expert referees to the professional and scientific standards expected of a proceedings journal published by IOP Publishing.

- **Type of peer review:**

For Peer review of submissions was used open system. Based on the results of the speeches, the editors of the Conference selected the most interesting scientific reports and proposed to publish the article in the Proceedings.

- **Conference submission management system:**

The organizing committee's email address was used to submit applications for the conference.

- **Number of submissions received:**

The organizing committee received 91 submissions.

- **Number of submissions sent for review:**

72 submissions were sent for review.

- **Number of submissions accepted:**

29 submissions were accepted by Editors.

- **Acceptance Rate (Number of Submissions Accepted / Number of Submissions Received X 100):**

32.

- **Average number of reviews per paper:**

One review for paper was prepared.

- **Total number of reviewers involved:**

24 reviewers worked with applications.

- **Any additional info on review process:**

- **Contact person for queries:**
Ph.D. Maksim Prozhega
Bardina, 4, 119334, Moscow, Russia

prmaksim@gmail.com

IOP Conf. Series: Materials Science and Engineering 1023 (2021) 012001 doi:10.1088/1757-899X/1023/1/012001

Peculiarities of the images of microscroches during load testing for fatigue fracture

R S Akhmetkhanov

Mechanical Engineering Research Institute of the Russian Academy of Sciences,4, Maly Kharitonyevsky Pereulok, Moscow, 101990, Russia

mibsts@mail.ru

Abstract. The paper investigates the features of changes in the fractal dimension of microsection images during load tests for fatigue life. For this, the methods of the theory of fractals were used - the assessment of the fractal dimension of the images of microsections at different stages of loading, and their multifractal spectra were also determined. It was found that before the destruction of the sample, the multifractal spectrum degenerates into a monofractal.

1. Introduction

The safety of technical systems depends on many factors. One of them is the mechanical strength of materials - the ability to withstand functional loads with a given margin of safety. At the same time, the creation of new materials with specified enhanced mechanical properties and the determination of the relationship between the structure (structure) of a substance and its physical and mechanical properties are among the most important tasks.

To ensure the safety of technical systems, various methods of non-destructive testing are used, while it would be important to obtain methods of visual inspection for special microsections using methods that allow this.

2. Methods

Currently, for the qualitative and quantitative analysis of the structural properties of materials, computer methods of information processing are used. Fractal analysis is one of the emerging trends in digital image processing. The development of this direction is facilitated by the fact that most images can to some extent be considered fractals or multifractals.

Fractal image estimation methods are used to study the structural properties of materials, fractures (study of the properties of the microrelief of structural materials) For example, in [1], the results of studies of the relationship between the fractal dimension of the surface of an object material and its mechanical properties (coefficient of friction, impact strength, dynamic modulus of elasticity, corrosion resistance, durability) are presented. A correlation was obtained between the mechanical properties of materials and the fractal dimension D of their surface.

The following works [2, 3] are devoted to the study of the correlation between the mechanical properties of materials and their fractal characteristics of the surface.

3. Results

To analyze the correlation between fatigue life and fractal dimension, an analysis of a series of microsections obtained in the study of 12GS steel (low-alloy structural steel for welded structures) was carried out during cyclic tensile stress tests. The microstructure of 12GS steel mainly consists of ferrite and pearlite. The structure of the material is granular, large grains of ferrite and small grains of pearlite.

When examining a microsection, its image can be presented in various forms - phases, boundaries, framework, etc. In this case, the fractal dimension can be determined in the following boundaries: $1 < D < 2$ and $2 < D < 3$ depending on the consideration of a flat image or presenting this image in three-dimensional form (taking into account the color of pixels).

It was found that up to 38000 loading cycles the image of a microsection remains almost unchanged (by the number of grains). During this cycle of loading, microcracks appear in the grains. The most intense increase in microcracks and their growth occurs in the range from 38000 to 65000 loading cycles, their accumulation occurs at approximately the same rate with a slight increase after 84000 loading cycles.

For the test of the sample with the limiting number of loading cycles N = 95000, in addition to the fractal dimension, other characteristics of the microsection image were calculated. With the accumulation of damage, the number of grains increases, while the characteristics of the area and perimeter of the grains decrease, i.e. the destruction of the grains occurs. The structure of the image becomes more heterogeneous. This is fixed in the image of the skeleton (figure 1 (b)). These images of the skeleton correspond to fractal dimensions $D_0 = 1.208$ and $D_{95000} = 1.253$. In this case, an increase in the index of fractal dimension indicates that the structure of the material has undergone changes - the appearance of a significant amount of micro-destruction (accumulation of damage occurs).

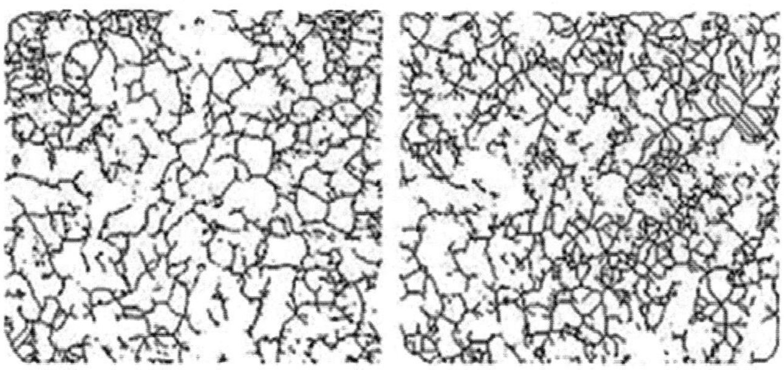

Figure 1. The skeleton of a microsection image at loads N = 0 (a) and N = 95000 (b).

In figure 2 shows examples of calculating the fractal dimension of the boundaries between phases.

The calculated fractal dimensions for the phases of the material and the boundaries between them showed that with an increase in the number of loading of the sample, the fractal dimensions of the image skeleton, the ferrite phases increase, and the fractal dimension for pearlite decreases. The fractal dimension of the boundary between the phases changes insignificantly.

The analysis of microsection images showed that the multifractal spectrum $f(\alpha)$ of the images changes to a monofractal spectrum after 61000 loading cycles. Also, with an increase in the number of loads N, the fractal estimate D_0 grows - from 1.6712 to 1.7485 (figure 3).

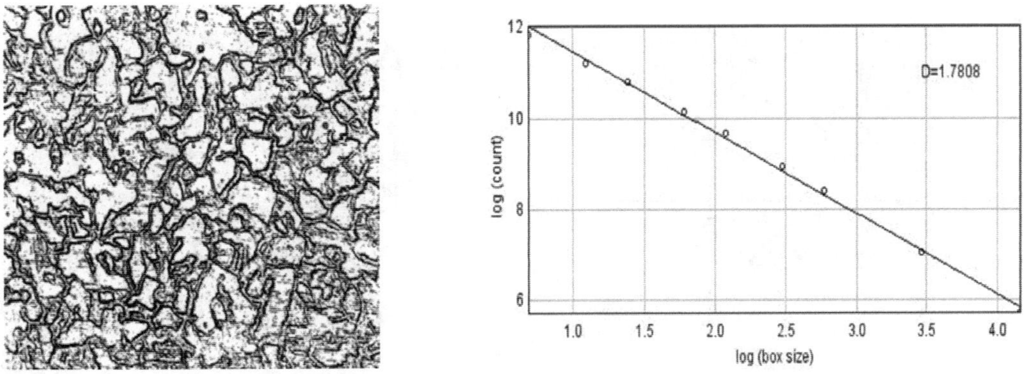

(a) (b)

Figure 2. Image of a microsection (selection of boundaries between phases and particles of ferrite and cementite in pearlite) and determination of the fractal dimension of this image.

The second series of the experiment continued up to N = 112000 loading cycles. The microsection with the number of loads N = 112000 also had a monofractal structure. The width of the multifractal spectrum S decreased, and the left branch of the spectrum also decreased. A significant change in the width of the multifractal S shows that the multifractal spectrum degenerates into a monofractal. The number of fractal dimensions decreases in a given multifractal spectrum.

To clarify the structure of the material, microsection images were clustered using the k-means method. The k-means method is a method of cluster analysis, the purpose of which is to divide m observations (from the space R^n) into k clusters, with each observation referring to the cluster (to the center centroid), who is close.

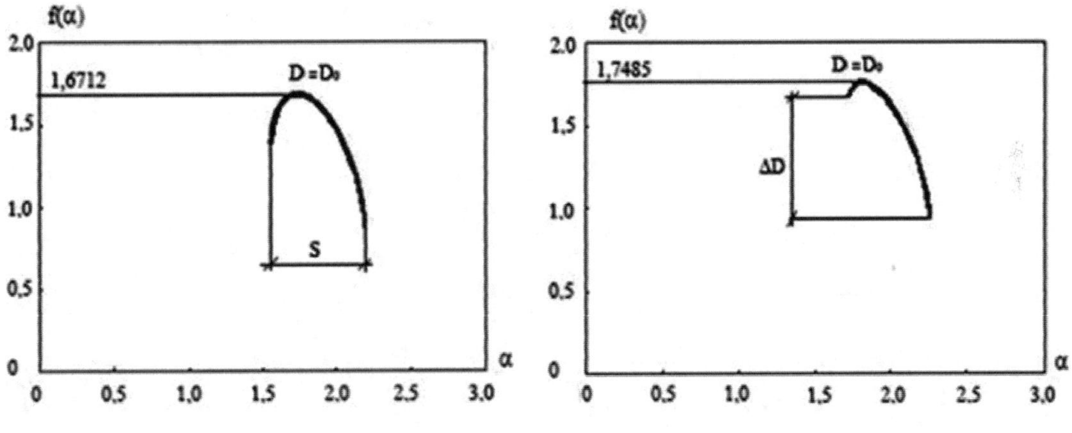

(a) (b)

Figure 3. Multifractal spectrum of a microsection image at the number of loads N = 0 (a) and N = 95000 (b).

When examining the images, the number of clusters was selected 10. In figure 4, they are highlighted in gray. If for the image of a microsection before testing the structure of the material had only 5 clusters, the division into clusters did not change the image of the microsection. Whereas before destruction, the microsection image already had 9 clusters. As you can see, cluster number 1 (the darkest areas) reflects small fracture zones of the sample material.

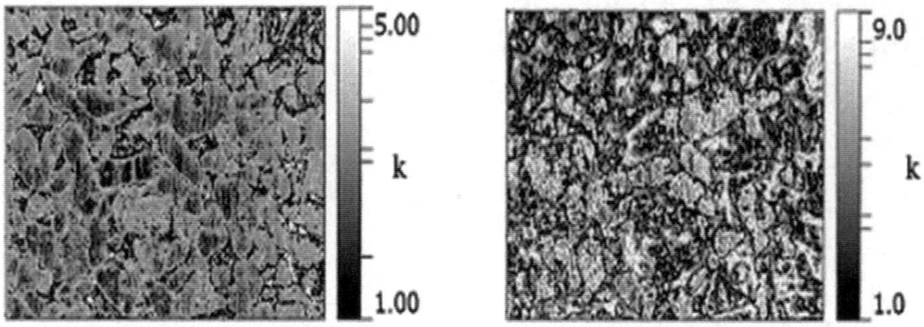

(a) (b)

Figure 4. The result of processing microsections using the k-means clustering method (10 clusters): (a) - with the number of loading cycles N = 0; (b) - with the number of loading cycles N = 112000

4. Conclusion

According to some scientific publications, it is known that in the absence of loading with an increase in the fractal dimension of the material structure, the strength increases, in the case of fatigue strength tests when micro- and macrocracks appear on the surface of the material, then an increase in the fractal dimension is the first visual sign of the impending destruction of the sample. And before destruction, the thin section acquires a monofractal distribution of the scale elements of the material structure, which corresponds to the criterion of the transition of the system from one state to another [4]. That is, the multifractal spectrumone changes - the number of fractal dimensions in the multifractal spectrum decreases, the spectrum width decreases - the multifractal spectrum degenerates into a monofractal one.

References

[1] Bovykin O B 2013 *Izvestia "MAMI"* **2** 14-8
[2] Ananyeva E S, Novikovskiy E A, Ananiev M I, Markin V B and Ishkov A V 2012 *Polzunovskii Vestnik* **1** 10-4
[3] Krasikova I E 2007 *Mathematical models and computational experiment in materials science* (Kiev: Inst Probl. Materials Science NAS Ukraine) **9** 79-84
[4] Akhmetkhanov R S 2019 *Problems of safety and emergency situations* **5** 20-9

Application of fractal theory in diagnostics of composite materials

R S Akhmetkhanov

Mechanical Engineering Research Institute of the Russian Academy of Sciences,4, Maly Kharitonyevsky Pereulok, Moscow, 101990, Russia

mibsts@mail.ru

Abstract. The article presents the results of studies to identify the features of diagnostic acoustic signals when defects are detected in composite materials using the methods of the theory of fractals. The use of the analysis of the spectral power density (PSD) and wavelet spectra of these signals did not lead to the detection of significant quantitative differences in the acoustic signals during the process of sample diagnostics (panel 4 mm thick). Acoustic signals from the standpoint of analyzing the characteristics of multifractal spectra were analyzed. It is shown that the defect state of the sample corresponds to the monofractal characteristic of the multifractal spectrum.

1. Introduction

The safety of technical systems depends on many factors. One of them is technical diagnostics. To ensure the effectiveness of diagnostics, it is necessary to develop computer methods for studying the obtained diagnostic data. The widespread use of composite materials that have their own distinctive mechanical and structural characteristics from metals and alloys requires the development of non-destructive testing methods that take into account these features. Therefore, the task was set to study the possibility of using the methods of the theory of fractals for analyzing diagnostic data and obtaining criteria for distinguishing acoustic diagnostic signals with similar characteristics of the signal power distribution density (PSD), which is often required when determining defects in a composite material during diagnostics by acoustic methods.

2. Methods

The purpose of this work is to study the features of acoustic signals in the diagnosis and detection of defects in composite materials, to determine the most sensitive numerical parameters that depend on the presence of defects.

In experiments on diagnostics of damage in a sample made of composite material (CM), a low-frequency acoustic AD-701 flaw detector was used, designed for non-destructive testing of multilayer structures and products made of laminated plastics, both in production and operation conditions. Acoustic control was carried out at 5 points of the sample - one control point (1) and 4 control points, where the amplitude values of the diagnostic signal were higher than the control value. Sample - panel 4 mm thick, 600 mm wide, 800 mm high.

The impedance method that was used makes it possible to detect defects such as delamination and non-gluing in non-metallic coatings and products made of laminated plastics, lying at a depth of 15-20 mm. This method is used to control joints that have several layers: to detect discontinuities in adhesive and soldered joints, to control the quality of the fit of studs, pins, axles and other parts installed with an interference fit. This control does not require bilateral access to the product and the application of a couplant on its surface [1,2].

3. Results

In figure 1 shows the wavelet spectra C(f,n) of acoustic signals measured at control points 1, 2 and 4 [3]. It is difficult to evaluate them and find differences in their wavelet spectra. You can give the qualitative characteristics of the frequency spectrum, their variability. But nothing more.

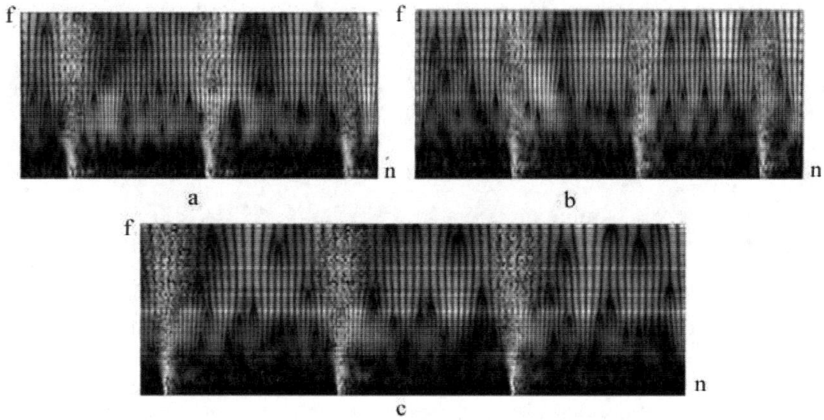

Figure 1. Wavelet spectra of acoustic diagnostic signals at control points: 1 (a), 2 (b) and 4 (c).

The clustering of wavelet spectra C(f,n) of acoustic signals by the k-means method at all diagnosed points did not show any difference in the number of clusters. The number of clusters in all signals was 10. A change in the number of clusters would indicate the presence of differences in the structure of the signal.

Also, for the wavelet spectra of acoustic signals, the Minkowski coupling function χ was calculated, which quantitatively estimates the coupling of the wavelet spectrum image C(f,n) by the values of the energy E of the frequency components in the signal

$$\chi = \frac{E_W - E_B}{n},$$

where E_W, E_B - symbols of the energy values of the frequency components in relation to a certain energy level E, which are determined from the following condition: $E_W < E$, $E_B > E$; n is the number of pixels in the wavelet spectrum image.

Figure 2 shows the Minkowski connectivity functions for points 1 and 4. The difference between these functions is that for point 1 there is more significant connectivity (+, -) at the level of values close to $10 \cdot 10^{-3}$, while at point 4 it is less - $7 \cdot 10^{-3}$.

It is known that during the transition of systems from one state to another, i.e. as the bifurcation point is approached, the susceptibility to noise increases in the system, with soft modes making the main contribution. The faint noise inherent in any real system is amplified. Small low-frequency perturbations become fundamental modes (order parameters), and high-frequency ones become subordinate to them. The growth of soft modes near the bifurcation of the stationary regime is universal. Thus, one can judge about the approach of a bifurcation (catastrophe) in advance by the growth of soft modes in the spectrum of its noise [4].

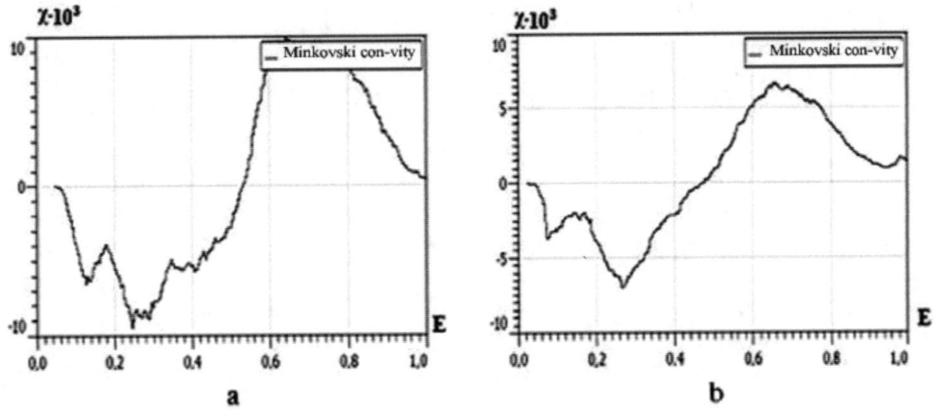

Figure 2. Minkowski connectivity χ for control points: (a) - 1; (b) – 4.

The resulting frequency response near the bifurcation can be described in the form of the dependence $S(f) \approx f^{-\beta}$, which describes power-law self-similar laws. The power spectra (PSD) (squares of amplitudes of the Fourier transform), often called noise, are especially adherent to simple homogeneous power laws of the $f^{-\beta}$ type. Among the noise, white noise with a spectral index ($\beta = 0$) is widely known. Depending on the value of the index β, noise is distinguished: white, pink (f^{-1}), brown (f^{-2}) and black (f^{-3}) [5]. This type of power-law dependence of the spectral characteristic is monofractal - a term used in the theory of fractals.

In this article, it was checked how the acoustic signal changes in the defect zone according to the type of the multifractal spectrum. Monofractal and multifractal dynamic processes differ in the nature of the power spectral density distributions. If in a monofractal process the distribution of the spectral power density can be described by a function that depends on the frequency f in the form - $S(f) \approx f^{-\beta}$, where β is a constant, then in the case of a multifractal process, the PSDs are described by a more complex dependence.

As noted above, a composite specimen was used in defect detection studies. According to the results of the inspection of this panel, 4 points were found (points 2-5, see the table), where the impedance amplitude was higher than the set threshold for point 1.

The multifractality of the process is usually represented by a multifractal spectrum (singularity spectrum) f(α). Multifractal spectra are characterized by spectral width, asymmetry, curvature [6].

The multifractal spectrum of (singularities) f(α) characterizes the dependence of the number of elements of the covering N_ε with different scales ε corresponding to points with the singularity exponent equal to some value

$$N_\alpha(\varepsilon) \sim \varepsilon^{-f(\alpha)}$$

In terms of meaning, the value f(α) under the condition α = const corresponds to the Hausdorff dimension. In this case of uniform distribution of the measure on the set, the spectrum of singularities is a single point on the plane (α,f), which corresponds to a monofractal process. With a nonuniform distribution of the measure, the function f(α) has a more complex (bell-shaped) form. The function f (α) is called the multifractal spectrum, characterized by the values D_i ($i = -\infty,\ldots,2,1,0,1,2,\ldots,\infty$), the asymmetry and spectrum width S. A decrease in the spectrum width S leads to the loss of multifractality. When the number of significant fractal dimensions D_i in the spectrum becomes minimal, it becomes monofractal.

Since the PSD of acoustic signals were visually similar, their multifractal spectra are similar, with the difference that for the defect-free zone, the fractal dimension is determined by the value $D_0 = 1.6090$ (the maximum point on the multifractal spectrum, figure 3), and for the defect-free zone value D_0 = 1.6356 (see table 1). These values are close and cannot be an objective criterion for detecting a defect in a research object.

Table 1. Data of acoustic signals and their multi-fractal spectra.

Number of point	Maximum signal amplitude A_m	Fractal dimension D_0	Inhomogeneity index spectrum ΔD	Width of the multi-fractal spectrum S
1	0.037	1.6090	0.609	0.975
2	0.038	1.4732	0.423	1.065
3	0.148	1.4957	0.745	0.950
4	0.295	1.6356	0.273	0.715
5	0.140	1.6232	0.550	1.435

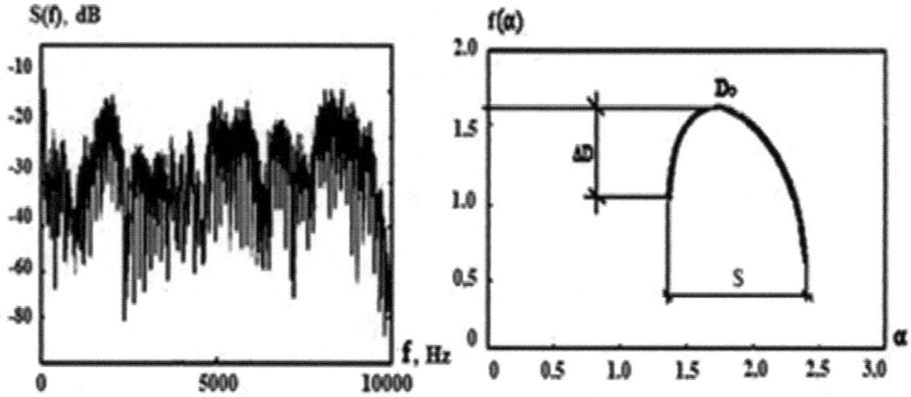

(a) (b)

Figure 3. PSD of the acoustic signal: at point 1 (a) and its multifractal spectrum (b).

It turned out that if we compare two points (1 and 4, table 1), they similar in shape of the PSD and to the histograms of the distribution of acoustic signal amplitudes, but they differ significantly in the multifractal spectrum. For point 4 the multifractal spectrum degenerates into a monofractal.

4. Conclusion

Thus, the presented results show that the type of multifractal spectrum is related to the presence of a defect or significant inhomogeneity of the material. Defective structures correspond to signals with a monofractal characteristic of the multifractal spectrum, which corresponds to the predominance of conventionally low-frequency components in the signal in relation to high-frequency components in a certain frequency range.

References

[1] Yakubovich B I 2016 *Uspekhi prikladnoi fiziki* **4** 127-38
[2] Razumenko D 2008 *Components and technologies* **9** 168-74
[3] Dyakonov V 2002 *Wavelets. From theory to practice* (Moscow: SOLON-R) 448
[4] Haken G 2005 *Information and self-organization. Macroscopic approach to complex systems* (Moscow: KomKniga) 248
[5] Rumanov E N 2013 *Uspekhi fizicheskikh nauk* **1** 103-12
[6] Akhmetkhanov R S 2018 *J. of Machinery Manufacture and Reliability* **47** 235-40

Fracture of a gas pipeline manufactured by high frequency current welding

A A Alexeev, A S Syromyatnikova, A M Bolshakov and A R Ivanov

Larionov Institute of the Physical-Technical Problems of the North SB RAS, Yakutsk, Russian Federation

spinor03@gmail.com

Abstract. Fracture of an underground gas main with a diameter of 530 mm during the production of pneumatic tests along a factory longitudinal welded joint made by contact welding using high-frequency currents, is considered. Fracture research was carried out by analysis of the project and technical documentation of the object, visual and measuring control of the object, ultrasonic measurement of pipe wall thickness, fractographic, metallographic and chemical analyses, measurements of hardness and microhardness of pipe metal. It was established that fracture of the gas pipeline occurred as a result of the opening of one pipe along the entire length of the factory-made longitudinal welded joint during the propagation of cracks in the welded joint and the subsequent propagation of the main crack in both directions to adjacent pipes; the centers of fracture were numerous surface crack-like defects of the welded joint. It is shown that a fracture was caused by a violation of the welding technology of a longitudinal welded joint, a violation of heat treatment, an unacceptable thinning of the pipe wall in the weld zone.

1. Introduction

Ensuring the safety of oil and gas pipelines, equipment of the oil and gas industry is of great importance in the polar regions of Russia - the Arctic, where the elimination of the consequences of the accident requires significant material and time resources, in the full dependence of the life of the Arctic settlements on energy resources [1, 2, 3, 4].

The climatic conditions of the polar regions of Russia are characterized by low climatic temperatures and sharp continentality. The temperature period below 0°C is about 210 days and the minimum temperature reaches 60°C below zero, the difference in average temperatures is 100°C. Most of the territory belongs to the permafrost zones [3, 4, 5].

The causes of emergencies in pipelines, equipment of the oil and gas industry, their development into accidents and disasters are, as a rule, failures of technical systems due to design errors, violation of manufacturing technology, operating conditions and operating modes, degradation of material properties during long-term operation, natural phenomena such as earthquakes, tsunamis, etc. [1, 2]. The article provides an analysis of the causes of the catastrophic fracture of the main gas pipeline on the territory of the Republic of Sakha (Yakutia), as a result of violations of manufacturing and welding technology.

2. Equipment and materials

The analysis of the causes of fracture is based on the analysis of the design and technical documentation of the object, visual and measuring control of the object, ultrasonic measurement of pipe wall thickness, fractographic, metallographic and chemical analyzes, measurements of the hardness and microhardness of the pipe metal.

Fractographic analysis was carried out at the macro and micro levels using a set of visual measuring control and scanning electron microscope Philips XL 20 to identify the mechanisms of damage and destruction. Metallographic analysis was carried out on a Philips XL 20 scanning electron microscope to reveal the microstructure of the material. Spectral chemical analysis was carried out in order to determine the grade of steel using the Spectroport F instrument.

The wall thickness was measured by the VZLET-UT ultrasonic thickness gauge over sections in four planes. The hardness of the material was determined using the TEMP-2 portable hardness tester in places of thickness measurement.

3. Results and discussion

In the Republic of Sakha (Yakutia), in the winter of 2006, a section of the underground gas pipeline was destroyed during pneumatic tests with a pressure of P=5.6 MPa (design working pressure P=5.5 MPa) at an ambient temperature of 42°C, which was accompanied by soil discharge and gas pipeline opening. A general view of the destroyed section of the gas pipeline is shown in figure 1.

Figure 1. General view of fracture of the gas pipeline.

The underground gas pipeline consist of steel electric-welded straight-seam pipes with an outer diameter of 530 mm and a wall thickness of 7 mm, the pipes are connected by resistance welding using high-frequency currents. The calculation of ring stresses showed that at the time of fracture, stresses did not exceed the permissible values and met the condition $\sigma \leq 0{,}75\sigma_T$ according to RD 12-411-01(Instructions for diagnosing the technical condition of underground steel pipelines) and Building code 2.05.06-85 (Trunk pipelines). According to the design documentation, chemical composition and hardness, the metal of the pipes corresponded to the steel grade 09G2S (GOST 19281-89).

The destroyed section consists of three pipes welded by two mounting ring joints (figure 2). The middle pipe was destroyed by the movement of the main crack along the axis of the longitudinal factory welded joint rectilinearly with the opening along the entire length of the joint; at a distance of 530 mm along the gas direction from the mounting ring joint, a side crack 100 mm long, perpendicular to the fracture line, was found in the pipe body. The maximum opening of the lateral crack is observed at the factory longitudinal welded joint, and the radial scars on the open surface of the crack are normal to the pipe walls. These facts indicate that a lateral crack arose due to the opening of the pipe along the factory welded joint and propagated according to the transverse shear pattern. The fracture of the first and third pipes along the gas direction — cracks with a curved path that passed along the base metal and looped before stopping — was a continuation of the fracture of the middle pipe.

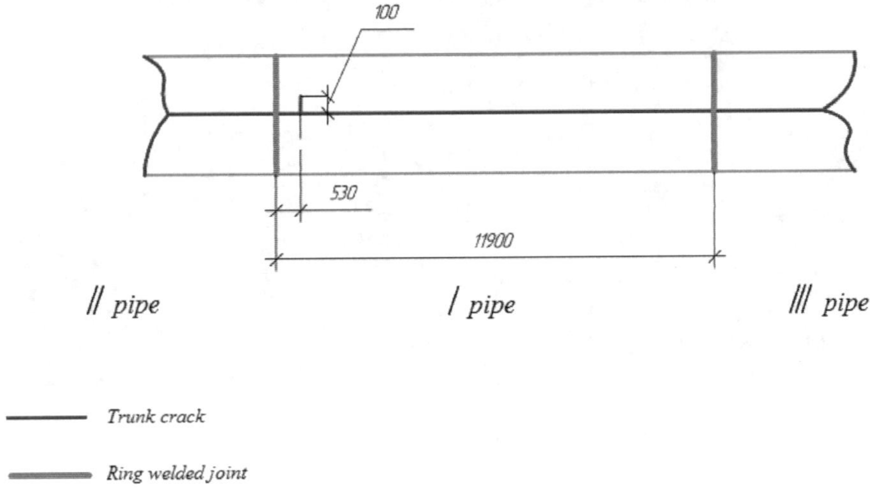

Figure 2. The scheme of fracture of the gas pipeline.

Far from the crack kink, an unacceptable thinning of the walls of the gas pipeline was not found, but almost all the measurements of the kink thickness do not meet the requirements of TU 14-3R-1471-2002, SP 101-34-96 (Set of rules. The choice of pipes for the construction of trunk pipelines): the pipe walls have a thinning beyond the minus tolerance, which not associated with local plastic deformation during fracture, and is the result of thinning of the pipe wall in the weld zone when removing the inner burr.

Microstructural studies of a thin section made along a plane normal to the surface being welded and to the pipe wall were carried out (etching 5% HNO_3 in ethanol using a Philips XL-20 scanning electron microscope). Three zones are distinguished by structure: the zone of the weld, the zone of thermal influence and the base metal.

The base metal has a ferrite-pearlite structure, fine-grained, a grain score of 10-11 and contains inclusions whose sizes are comparable to the grain score; perlite stitching along the direction perpendicular to the axis of the pipe. The weld zone with a width of 0.75 mm has a hardening structure and consists of two areas. The first region with a width of 0.5 mm has a martensitic structure, the second, transitional to the zone of thermal influence, an area with a width of 0.25 mm is an inhomogeneous structure consisting of martensite and ferrite precipitates. The thermal influence zone with a width of 0.5 mm has a ferrite-pearlite, fine-grained structure with an average grain size of about 5 μm, which is probably due to the fine-grained initial austenite and the occurrence of ferrite-pearlite transformation under conditions of overheating. Thus, in the zone of the welded joint of the middle pipe, a martensitic structure has formed, which causes a tendency in the welded joint to form brittle quenching cracks.

Upon visual inspection of the fracture surface of the middle pipe, three types of areas are distinguished (figure 3 (a)). Microfractographic analysis of these areas was carried out using a Philips XL-20 scanning electron microscope in the secondary electron mode.

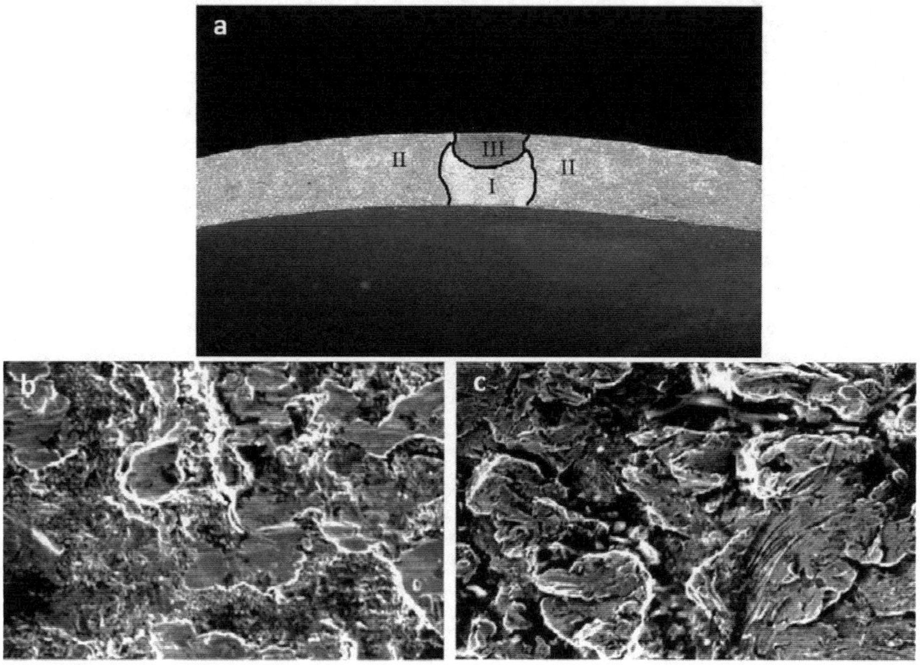

Figure 3. Fracture surfaces of the middle pipe: (a) general view; (b) type I area; (c) type II area.

Areas of type I, 2-10 mm wide, are characterized by a brilliant crystalline fracture, a fibrous relief, occupy all or part of the thickness of the destroyed pipe surface, are distributed irregularly along the entire length of the destroyed surface, and are fibrous zones of the fracture surface corresponding to regions of slow growth of cracks from the fracture centers. The fracture surface in type I area consists of flat facets of brittle cleavage along martensitic plates, whose orientation is determined by the direction of the acting maximum stresses (figure 3 (b)).

Type II areas are characterized by a gray crystalline fracture, a rough relief, occupy the main part of the fracture surface and represent the radial zones of the fracture surface corresponding to regions of rapid crack propagation (figure 3 (c)). Radial scars indicating the direction of propagation of the crack emanate from the fibrous zones. The large number of fibrous zones (type I area) on the fracture surface and the multidirectional radial scars in various type II areas indicate the propagation of cracks from many centers distributed along the length of the destroyed surface.

Areas of type III - visually dark, covered with an oxide film, sickle-shaped, with a fibrous relief, are found in 28 areas of type I, located at the outer or inner surface of the pipe. The fracture surface is characterized by a quasi-split fracture characteristic of brittle fracture of quenching structures. Fibrous circular grooves in areas III extend from the linear portion of the free surface of the weld joint and extend concentrically deep into the weld joint. Oxidation of the surfaces of open cracks, as a rule, occurs when they form until the weld joint have completely cooled after welding. These facts indicate that the centers of fracture were surface crack-like defects of the welded joint that formed before the welded joint had completely cooled after welding.

4. Conclusion

Thus, the fracture of the gas pipeline occurred as a result of the opening of one pipe along the entire length of the factory-made longitudinal welded joint during the propagation of cracks in the welded joint and the subsequent propagation of the main crack in both directions to adjacent pipes; the centers of fracture were numerous surface crack-like defects of the welded joint. The pipeline fracture was caused by the following violations: a violation of the welding technology of a longitudinal welded joint, as indicated by the presence of crack-like defects in the welded joint formed in it until complete cooling after welding; a violation of heat treatment, as indicated by the formation of quenching structures; an unacceptable thinning of the pipe wall in the weld zone when removing the inner burr.

References

[1] Mokrousov S 2006 *Occup. safety in industry* **9** 16-19
[2] Smirnov A, Gerike B and Murav'yev V 2003 *Diagnosis of technical devices of hazardous production facilities* (Novosibirsk: Nauka) p 244
[3] Lyglayev A, Levin A, Kornev I, Cheremkin M and Bol'shakov A 2001 *Gas industry* **8** 37
[4] Zakharova M, Makhutov N, Lebedev M and Bol'shakov A 2017 *Bull. of Russ. Acad. Sc.* **87** **9** 858-62
[5] Alexeev A, Sypomyatnikova A, Levin A and Lyglayev A 2009 *Def. and fract. of materials* **2** 25-30

The effect of process parameters of gas-dynamic spray on generation of brass-type coating structure

V E Arkhipov, G V Moskvitin and M S Pugachev

Mechanical Engineering Research Institute of the Russian Academy of Sciences, Moscow, 101990, Russia

GVMoskvitin@yandex.ru

Abstract. The study results have shown the effect of coating temperature and contact ratio on generation of phase structure of the coating based on the mixture of Cu and Zn particles and Al_2O_3 applied by cold gas-dynamic spray method. The x-ray diffraction and micro-spectral analyses have shown the presence of copper and zinc in the structure and generation of electronic compound based on $CuZn_3$ (ε – phase) and Cu_5Zn_8 (γ – phase), which is preconditioned by copper diffusion into zinc. The mass fraction of intermetallic compositions in the coating structure depends on the coating temperature and contact ratio, it reaches 12% and 33%. The analysis of surface processing by solid and plastic particles flow has shown the effect of copper particles deformation on diffusion rate.

1. Introduction

Gas-dynamic spray coating is the advanced method of obtaining high-quality coatings in adverse conditions. The sprayed copper and zinc-based layer has high adhesion strength of not lower than 50 MPa and cohesion strength of 80 to 150 MPa, whereas the temperature of steel backing heating does not exceed 140°C.

The objective of this work is to study the changes in the brass coating structure depending on process parameters of gas-dynamic coating.

2. Methods and modes of obtaining the coating

The coating was sprayed on the backing of 5140H steel using gas-dynamic device DIMET-404 at air temperature 270°C, 360°C and 450°C, the rate of replaceable nozzle displacement in relation to the sample surface 10 $mm \cdot s^{-1}$, the distance from the nozzle cut to the surface equal to 10 mm and nozzle displacement of 2 mm (contact ratio 94%) and 3 mm (contact ratio 55%) (table 1). To apply the coating, the authors used the mechanical mixture of copper and zinc particles and aluminum oxide (Al_2O_3) of as received grade C-01-11 with ingredients weight ratio $Cu:Zn:Al_2O_3$ = 35%:35%:30%, manufactured at Obninsk Powder Spray Center [1].

Table 1. The results of particle surface treatment analysis.

Nozzle displacement, mm	Contact ratio,%	Number of particle surface treatment, times		
		1 (time/surface)	2 (time/surface)	3 (time/surface)
2	64	14 s / 20 mm^2	28 s / 90 mm^2	42 s / 112.5 mm^2
3	55	12 s / 74 mm^2	24 s / 148.5 mm^2	-

3. Research results

Chemical analysis of the coating composition has shown a significant dependence of the components content on the spray temperature. Rising the air flow temperature from 270°C to 360°C and 450°C was accompanied by generation of a coating where zinc mass fraction increases from 8.15% to 22.3% and 37.6%. The copper content in the coating composition at gradual air flow temperature increase decreased from c 91.7% to 77.5% and 62.2%.

The metals (copper, zinc), aluminum oxide and solid electron-type solution based on $CuZn_3$ (ε – phase) (table 2) were revealed in the coating applied at the air flow temperature of 270°C. Copper-zinc state diagram shows that ε and ε+η – phases were generated in case of zinc presence in the melt in the amount of ≥ 78 at.% [2]. The increase of nozzle displacement distance from 2 mm ($K_{cont.}$ 64%) to 3 mm ($K_{cont.}$ 55%) resulted in copper content reduction from 87.9% to 83.8%, which could be explained by the increase of weight ratio of ε – phase, which includes copper (table 2). The study of the coating applied using the mechanical mixture of copper, zinc and aluminum oxide particles by gas-dynamic coating reveals the presence of the process of copper into zinc diffusion with generation of electron type compound. However, many researchers of diffusion process in copper-zinc system have theoretically and practically proven that zinc into copper diffusion coefficient exceeds the copper into zinc diffusion coefficient by several times; sometimes this ratio, for example, calculated for ε – phase, is of abnormal nature ($D_{Zn} : D_{Cu} = 47$) [3, 4]. In this case we could suppose the presence of other conditions that affect the diffusion process at gas dynamic spray of the coatings.

Table 2. Coating phase analysis.

T°C	Mass fraction of metals, phases and compounds, %							
	Cu		Zn		ε - phase		γ - phase	
	Nozzle displacement, mm							
	2	3	2	3	2	3	2	3
270	87.9±1.0	83.8±0.8	1.44±0.17	1.3±0.1	5.3±0.9	8.5±0.9	-	-
360	62.0±0.9	69.0±0.7	7.9±0.6	12.6±0.6	7.9±0.8	10.4±0.5	17.0±2.0	-
450	40.5±0.8	56.0±0.8	9.0±0.6	24.0±0.6	11.4±0.7	12.0±0.7	33.0±0.2	-

Rising the spray temperature to 360° C is accompanied by the major decrease in the mass fraction of copper and the increase of zinc and the intermetallic compound in ε - phase (table 2). When the coating is sprayed with the contact ratio of 64% (2 mm displacement), a new phase is revealed in the coating structure – the electronic-type solid solution based on Cu_5Zn_8 (γ - phase), with mass fraction of 17.0% (table 2). When smelting brass, this phase is generated when zinc content in the solution is within the range of 52-78 at.% (γ + β′; γ; γ + ε) [2]. Analyzing the ratio of metals and compounds in the coating with an increase in contact ratio, it is possible to observe that the decrease of pure copper content from 69.0% to 62.0% and zinc from 12.6% to 7.9% is due to the generation of a new γ - phase (Cu_5Zn_8) compound. The mass fraction of ε - phase decreases as well, which is also associated with the process of copper into zinc diffusion and the generation of a new electronic compound.

The increase in the content of $CuZn_3$ (ε-phase)-based electronic compound in the structure of the coating composition with the increase in the spraying temperature at a nozzle displacement by 2 mm from 5.3% to 7.9% and a nozzle displacement by 3 mm from 8.5% to 10.4 % is due to the increase of the spraying temperature [5]. Moreover, the increase in the content of the electronic phase (ε-phase) in both versions of gas-dynamic spray slightly differs by 2.6% and 1.9%

The next increase in the spray temperature to 450°C is accompanied by generation of a coating, in which the amount of copper significantly reduces, and the mass fraction of zinc increases, which is due to the change in the metal content at spray temperature increase (table 2). The mass fraction of $CuZn_3$-based electronic compounds (ε-phase) and Cu_5Zn_8-based electronic compounds (γ - phase) increases, which is related to the increase in the diffusion rate at increasing spray temperature.

Decreasing of the distance by which the nozzle is displaced when the coating is sprayed on the entire surface of the sample (3 → 2mm) results in significant copper (56.0% → 40.5%) and zinc (24.0% → 9.0%) content reduction which is associated with generation of gamma phase in an amount of 33%. At a spray temperature of 450°C, the change of nozzle displacement distance does not result in significant change in the ε-phase content (table 2).

The results of X-ray diffraction phase analysis show that when the coating was sprayed at air flow temperature of 360°C and 450°C, γ-phase was formed in the applied metal layer in an amount of 17% and 33%. And when the nozzle was displaced by 3 mm, at equivalent other spray parameters, this intermetallic compound was absent in the applied coating. In this case, the revealed trend indicates the presence of the effect of gas-dynamic spray parameters, in particular, the cover ratio, on the phase composition of the coating. A graphical assessment of processing options reveals that in the first case (2 mm) there are surface areas (S≈50%) exposed to the particle flow for the third time (table 1).

In this case, the observed effect of the diffusion process intensity increase with the generation of a Cu_5Zn_8 –based electronic compound (γ - phase) can be related only to the additional deformation of copper particles. In this case, deformation can change the shape (geometry) of the particle, which should be accompanied by the displacement of the boundaries of contact between copper and zinc particles, but not lead to the acceleration of the diffusion process. In case of generation of stresses, their relaxation will occur due to elementary acts of cooperative copper atoms displacement (jumps), which will accordingly affect the diffusion coefficient and rate [5].

Besides, the calculations have shown that in the process of spraying, the temperature of the applied metal coating in the area of direct air flow effect was 80-90% of the gas stream temperature, that is, 290-320°C and 360-405°C when the coating is applied at the temperature of 360°C and 450°C [6]. When zinc was heated to the temperature close to the melt temperature (≈419°C), the number of vacancies significantly rose, which provided more intense copper to zinc diffusion process according to the vacancy principle. Therefore, copper to zinc diffusion process with the generation of electronic compounds inherent to brasses, occurs due to the copper particles deformation with corundum particles, generation of stresses and micro-deformations with subsequent deformation thereof due to cooperative copper atoms displacement (jumps) and significant increase on vacancies in zinc.

Microspectral analysis of coatings applied at the spray temperature of 360˚C and 450˚C and nozzle displacement by 3 mm, was performed along the route that began on the copper particle, further passed by zinc particle and finished on copper particle (figure 1 (a), (b)).

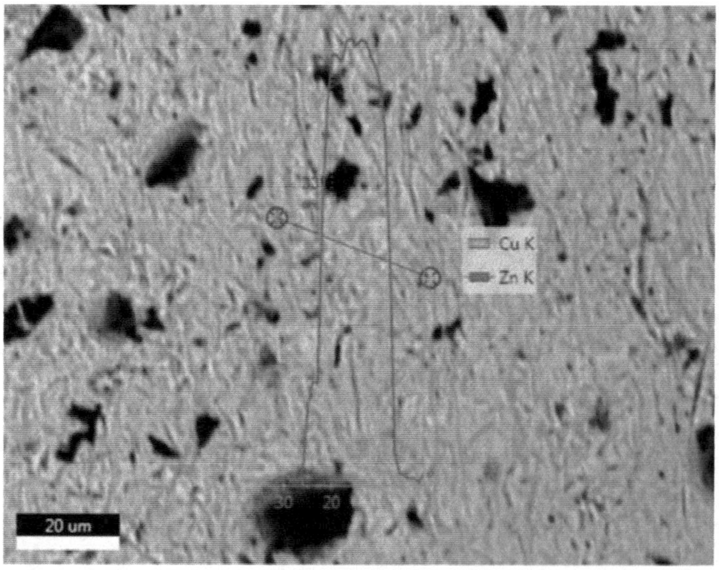

(a)

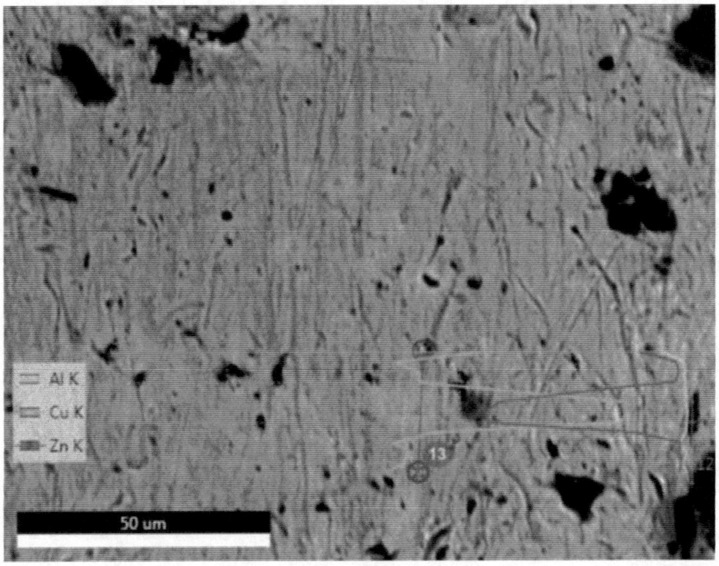

(b)

Figure 1. Copper and zinc particles coating applied at the temperature: (a) – 360˚C and (b) – 450˚C.

The analysis of copper and zinc distribution along the scanning route has shown the presence of copper diffusion into zinc at the both sides of zinc particle (figure 2). The results obtained show that in the process of spraying at air current temperature 360˚C there exist two areas with different copper and zinc ratio: a more significant one, where copper amount is in the range of 7-10 at.% (90-93 at.% of zinc), and a less significant one, where copper content is 18-20 at.% (80-82 at.% of zinc). Judging from copper-zinc state diagram, the electronic compound ε-phase was generated at this copper and zinc ratio, which confirms the results of x-ray diffraction phase analysis. Moreover, in this case we can speak about the presence of electronic type phase, where zinc and copper atoms ratio differs.

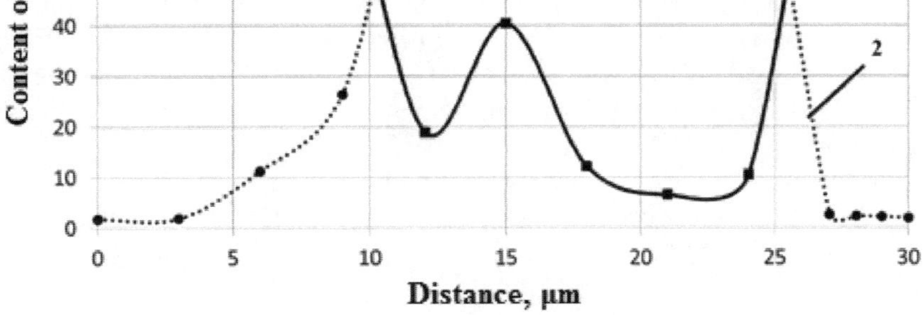

Figure 2. Copper and zinc distribution in the coating applied at the temperature of 360°C (1 – copper, 2 – zinc).

The rise of spray temperature (450 °C) was accompanied with generation of a coating, where and area with copper content of the order of 20 at.% and zinc content of ≅80 at.% (figure 3) is generated due to the copper diffusion into zinc particle. Copper-zinc state diagram shows that at this components content the ε-phase electronic compound was generated, which confirms the results of x-ray diffraction phase analysis (table 2). That is, the rise of spray temperature at equal other process parameters of spray accelerated the process of copper to zinc diffusion.

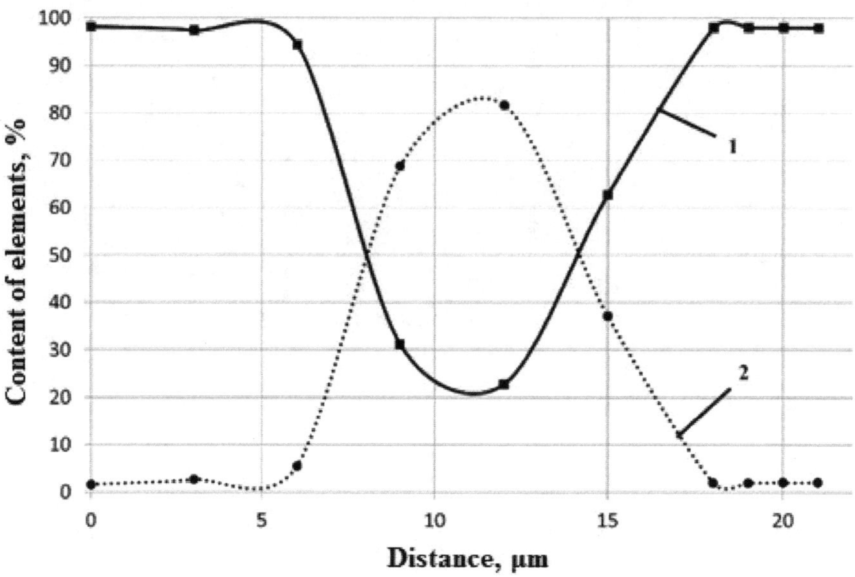

Figure 3. Copper and zinc distribution in the coating applied at the temperature of 450°C (1 – copper, 2 – zinc).

At mutual (reactive) diffusion, measuring the diffusion layer depth, it is possible to determine the diffusion coefficient using the formula [5]:

$$D = X^2 \cdot (2t)^{-1}, \quad (1)$$

where X is the mean diffusion layer depth (a half of diffusion layer depth is taken for calculation), m; t is the diffusion process time, s.

Scanning route length was 24 μm; the section, where the copper to zinc diffusion process occurred, did not exceed 14.0 μm ($14 \cdot 10^{-6}$ m). In this case, if we consider the time of the direct impact of the flow of particles and heated air on the coating with a nozzle displacement of 3 mm equal to 24 s (table 1), the copper into zinc diffusion coefficient was $\approx 0.11 \cdot 10^{-13}$ $m^2 \cdot s^{-1}$.

The coefficient of diffusion from liquid zinc melt into copper at the temperature of 427°C is $\approx 0.45 \cdot 10^{-13}$ $m^2 \cdot s^{-1}$. In another work, the coefficient of diffusion of zinc into copper for γ-phase at the temperature of 350°C was calculated, it equaled to ≈1.3x10-13 $m^2 \cdot s^{-1}$ [3, 4]. Therefore, the presented results obtained in this work, show that during gas-dynamic coating spray the coefficient of diffusion of copper into zinc is higher than that of zinc into copper from the melt.

Microspectral analysis was carried out on the coating applied at 450°C and nozzle displacement of 2 mm. From the area of two copper particles (spectra 1–6 and 14–19) and one zinc particle (spectra 7–13), we studied the coating components concentration distribution (figure 4).

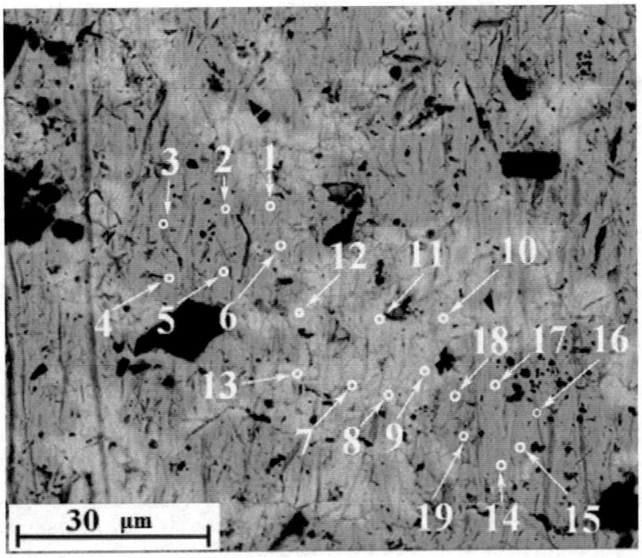

Figure 4. The analysis of the chemical composition of a coating surface after spraying at the temperature of 450°C (the areas of elemental analysis are highlighted).

The processing of the results for copper particles showed that the distribution of copper and zinc at 10 points was quite stable and their atomic ratio averaged to 97.59% and 2.41%. At two points, we observed an increase in zinc concentration up to ≈5.1%. Besides, we observed an abnormal change in the components concentration in copper particles, when at the distance of ≈ 3 µm the copper concentration varied from 39.11 at.% up to 98.45 at.% (figure 4, point 10 and point 11). This result can be interpreted as the presence of mechanical mixing of metals under the effect of corundum particles, which was noted in the study of a coating based on the mixture of aluminium and zinc particles [6]. In a zinc particle, the concentration of copper varies in a significant range from 13.26% to 52.86%. According to copper-zinc phase diagram, the regions with such zinc content corresponded to the existence of electronic type compounds of ε- and γ-phases, which is confirmed by the results of x-ray diffraction analysis (table 2).

4. Conclusions

An increase in the spray temperature from 270°C to 450°C is accompanied by a decrease of copper mass fraction from 91.7% to 62.2% and the increase of zinc mass fraction from 8.15% to 37.6% in a coating based on copper and zinc particles mixture.

It is shown that when the nozzle is displaced by 3 mm, the structure contains copper, zinc and a $CuZn_3$-based electronic compound (ε-phase), the mass fraction of which increases from 5.3% to 11.4% with the spray temperature increase from 270°C to 450°C.

It is shown that when the nozzle is displaced by 2 mm, alongside with copper, zinc and compound (ε-phase), a Cu_5Zn_8-based electronic compound (γ-phase) is generated in the structure, the mass fraction of which increases from 17% to 33.0% when the temperature of the air flow rises from 360°C to 450°C.

The analysis of the sample surface treatment with a heated air flow and a mechanical mixture of metal and corundum particles indicates the effect of copper particles deformation on the predominant diffusion of copper into zinc and the increase in the diffusion coefficient.

References

[1] Dymet Application of technology and equipment (electronic resource) Available at: http://www.dimet-r.narod.ru/ (accessed 23 August 2020)
[2] Efremov B N 2016 *BRASSES from phase structure to structure and properties*
[3] Zayt V 1958 *Diffusion in metals*
[4] Smiths K J 1980 *Metals: A Handbook*

[5] Bokshtein B S 1978 *Diffusion in metals*
[6] Arkhipov V E, Landarski A F, Moskvitin G V and Pugachev M S 2017 *Gas-dynamic coating: structure and properties of coatings*

IOP Conf. Series: Materials Science and Engineering 1023 (2021) 012005 doi:10.1088/1757-899X/1023/1/012005

Quality control of composites

A V Berezin[1], I S Maidanov[2], S A Smelov[2] and I V Gadolina[1]

[1]Mechanical Engineering Research Institute of the Russian Academy of Sciences (IMASH RAN), Moscow, Russia
[2] ORPE "Technologiya" named after A G Romashin, Obninsk, Russia

gadolina@mail.ru

Abstract. According to the kanban (just in time) system, any production system needs continuous improvement. Possessing excellent mechanical properties, composite materials demonstrate a significant variation of properties (in particular, strength indicators). The composite manufacturing process is multi-stage and requires optimization after thoughtful consideration of the nuances. Exploratory data analysis was applied as a first step for this analysis. It serves as a basis for further experimental research, which it is desirable to carry out using the mathematical theory of experimental design. After the first stage of the study of the process, namely, the study of the statistical reproducibility of the technological process, in order to obtain more complete information about the process, it will be necessary to introduce additional influences on it.

In the problem of quality control of polymer composite materials (PCM), modern control methods are used, including ultrasonic diagnostics. We consider the composites in which high-strength carbon materials (filler) are used as the main bearing elements, and a polymer thermosetting epoxy resin (matrix) is used as a binder. One should note that the quality of the indicator should be confirmed in a statistical aspect [1]. During control, researchers should measure several characteristics: compressive and tensile strengths, shear strength, Young's modulus, etc. In this study, we used exploratory data analysis [2]. Below are some of the results of the analysis results concerning quality control for selected lots of products.

Methods for data acquisition

Scattering fields

Figure 1 shows the scattering fields for some characteristics of one of the batches of composites manufactured in 2019. Were considered 1) density; 2) ultimate tensile strength; 3) ultimate compressive strength. The figure shows a positive correlation between the experimental tensile and compressive strengths, which is not surprising. Although it is theoretically known that an increase in density leads to an increase in strength, for this series of experiments the correlation is not entirely unambiguous, which can be explained by the insufficient number of experiments. However, the factor of increasing the density should be kept in mind in the future when planning the experiment.

IOP Conf. Series: Materials Science and Engineering 1023 (2021) 012005 doi:10.1088/1757-899X/1023/1/012005

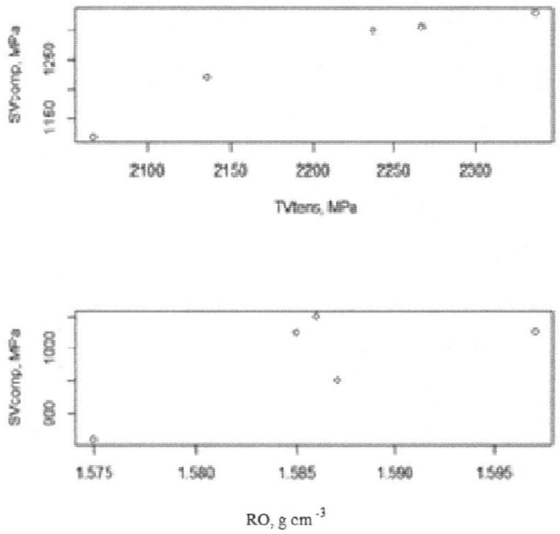

Figure 1. Scattering fields of one of the batches of composites.

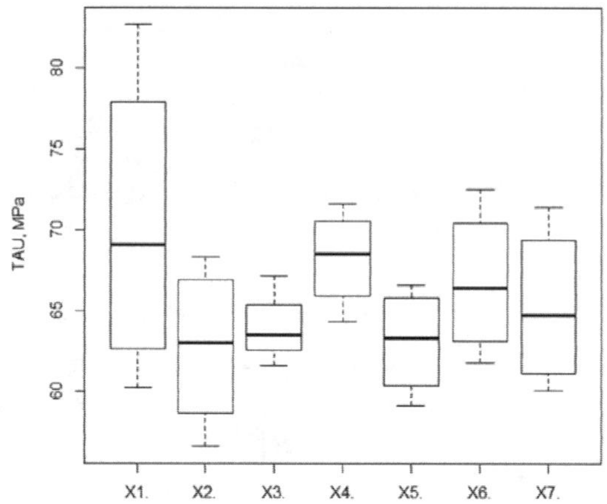

Figure 2. Outlier analysis boxplot chart.

Boxplot chart for outlier analysis

The authors of [3] previously noted that parametric methods for rejecting sharply distinguished values are volatile to deviations from distributions from normality. In this regard, the authors [3] used the nonparametric methods. Figure 2 shows a boxplot [4] chart as an example of one of the exploratory analysis tools. The value of 83.9 MPa in batch V1 is an outlier. That fact was also confirmed by the use of a particular statistical criterion. Its elimination allowed to increase the significance levels for all distributions.

Factors affecting the quality of physical and mechanical parameters of prepreg in the technological process

The list of qualitative and quantitative factors possibly affecting the strength of the test materials should be drawn up on the basis of expert opinion. The responses, that is, the functions to be optimized, are a material ultimate strength and its scatter. An approximate list of factors affecting the responses is shown in table 1.

Table 1. List of production factors.

Preparation of prepreg (factors influencing the stability of the technical process)	Sample making
Indoor temperature and humidity.	Layout of samples (manual labor).
The density of the binder in the bath and the temperature of the binder. Drying chamber temperature.	Autoclave forming of CFRP.
	Measuring slab thickness by points.
Shaft clearance.	Attaching the test pads.
Influence of raw materials (during storage) on the results.	
Influence of % binder content on the result (see table 2).	

Checklists

A checklist is the name of the form for collecting and registering data [6]. It is a tool designed to facilitate the collection of data and organize them to encourage further use (processing) of the collected information. A checklist is a paper form with pre-printed tables, documents, drawings for controlled parameters. Figure 3 shows examples of the checklists.

Checklist of types of sample defects			
Name ________	Party ________ —	Date ____________ —	Note (the tool)
Type of defect	**Inspection results**	**Number of defects**	
Crack			
Scratch			
Stratification			
Other			
		Subtotal:	

Checklist of types of sample defects			
Name ________	Party ________ —	Date ____________ —	Note the tool
Type of defect	**Inspection result**	**Number of defects**	
Crack			
Scratch			
Stratification			
Other			
		Subtotal:	

Measurements of the thickness of the sample/SI						
Nominal size	**Deviation (0.02)mm**	**Measurements**				
		1	2	3	4	5
***						

***-the boundaries of the tolerance field

Figure 3. Examples of the checklists during composite production.

Shown in figure 3 and similar examples can help the engineers to monitor the production, and in the future, built the control charts.

Management recommendations

Acceptance control paradoxes

Numerous debates about the appropriateness of incoming control are closely related to the problem of suppliers [5]. In Russia, the relationship between consumers and suppliers was based on suspicion, mistrust, and the search for the culprit. If we thoroughly check the raw materials (supplies) for passport data following GOST, it turns out:

- Rising prices for products. Waste of time for testing (7 days) and retesting in case of non-compliance with TU or GOST (7 days + double the number of samples).
- The comparison of measurements on initial samples with the results laid down by the manufacturer imposes specific requirements on the metrological characteristics of the supplier and the consumer on samples, materials, equipment.

Due to incoming inspection, people often waste time on the supplier's confirmation of the inherent characteristics. If the samples do not pass the tests the first time, then a second control is carried out on a doubled quantity. The incoming inspection increases the number of samples, increases the number of materials used for samples, increases the time of personnel involved in retesting, and requires additional equipment operation. Rejection of the principle is modern: more quantity, lower price. Retests are costly for the enterprise. The supplier must be reliable, and preferably, it must be the manufacturer for the possibility of adjusting the technical process.

IOP Conf. Series: Materials Science and Engineering 1023 (2021) 012005 doi:10.1088/1757-899X/1023/1/012005

Tables 2 and 3 show the results of the incoming inspection for a conditional batch of products No.1, carried out at the customer enterprise. Table 2 shows the analysis of a prepreg (preform for making a composite) impregnated with an epoxy binder. Besides, samples of batch No.1 were tested for the component composition of cured carbon fibre reinforced plastic, which is reflected in table 3.

Table 2. Analysis of prepreg (batch # 1).

№ Coil	Indicators		
	Volatile matter content, %	Binder content, %	Mass fraction of the soluble fraction of the binder, %
1	1.05	40.25	98.73
2	1.57	39.56	99.86
3	1.78	41.13	97.58
Requirements	< 3	38-42	< 95

Table 3. Component composition of cured CFRP.

№ n/n	Controlled indicators	Test results	Requirements
1	Apparent density $g \cdot cm^{-3}$	1.600	1.58-1.62
2	Porosity %	0.3	0.2-1.0
3	Volume fiber content %	67.2	60-70
4	Volume content of the binder %	32.5	30-40
5	Fiber mass content %	74.8	65-75
6	Mass content of binder %	25.2	25-35

Tables 2 and 3 show that the incoming inspection tests confirmed the compliance of all parameters of the coils of batch No.1 with the requirements, which means that the incoming inspection in this situation is redundant.

Discussion. The urgent steps

Metrology. Requirements for test pieces

An essential point in the analysis of experimental data is metrology. The authors of this article drew attention to the problem of the adequacy of the specimen's compression pads to certification tests. There is a wide range of problems, starting with the method of cutting samples (a diamond saw heats both filaments and binders, influencing the strength characteristics [7]), the design of the samples (mainly the specimen thickness and the number of layers), the material of the pads for testing, the method of fixing these patches. The current practice of the use of pins supporting the pads leads to the formation of a statically indeterminate problem, and the breaking stress characterizes an unknown quantity (either composite or pin shear strength).

In figure 4 the types of specimens breaking types are shown. Types (c) to (h) show the varied breaking types, but the samples are treated as a unit pool. The types of fractures (a) and (b) are excluded from the testing results, because in them the fracture took place in the pads. This raises the price for testing due to the increment of required specimens.

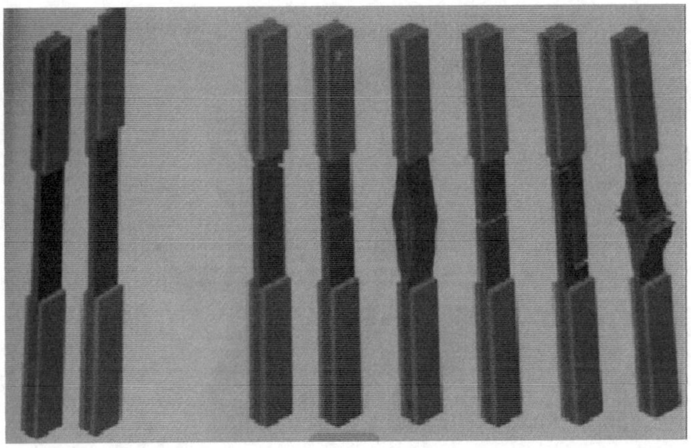

a b c d e f g h

Figure 4. Types of compression fracture in composite samples.

Table 4. Types of destruction of unidirectional samples.

a	b	c	d	e	f	g	h
Angled in the pads at the top of the specimen	Adhesive peeling of pads in the upper part of the sample	At an angle in the working area at the top of the sample	At an angle in the working area in the middle of the sample	Transversal lamination in the working area in the middle of specimen	Horizontal stratificatio n in the working area in the middle of the sample	Longitudinal delamination in the working area in the middle of the sample	Splitting in the working area in the middle of the sample
Not allowed		Allowed					

A large number of "spoiled" samples (type of break (a) and (b)) increase the cost of testing. This means that solving the problem of test metrology is a top priority.

Conclusions

The strength properties of composites depend on many factors and are subject to significant scatter in comparison with metals. Following the kanban system, a preliminary analysis of the production process was carried out. The use of exploratory analysis tools (scattering fields, a box with whiskers) made it possible to reveal some features of the strength indicators of composites. The task of improving the metrology of testing is discussed. Additional efforts should be directed to ensure that the test is performed to determine the important strength indicators - the ultimate compressive and tensile stress as well as shear testing under static conditions. Ultrasound testing also is important in quality control. Further analysis will involve the use of control charts to analyze common and specific causes of variation [2]. Under the principles of modern management, it is advisable to avoid (reduce) the incoming control whenever possible.

References

1. Berezin A V *et al* 2016 Application of CFRP composite materials in airframe units of civil aircraft of Tupolev PJSC *Problems of mechanical engineering and automation* **3** pp 4-9
2. Adler Y and Shper V 2019 *A practical guide to statistical process control* (Moscow: Alpina Publisher)
3. Lisachenko N G, Popov A G and Dumansky A M 2018 Decision making when calculating statistically valid characteristics of polymer composite materials at the stage of outliers analysis *Industrial laboratory Diagnostics of materials* **84(5)** pp 74-80

IOP Conf. Series: Materials Science and Engineering 1023 (2021) 012005 doi:10.1088/1757-899X/1023/1/012005

4. R Core Team 2020 A language and environment for statistical computing *R Foundation for Statistical Computing Vienna Austria*
5. Adler Yu P 2000 Love your suppliers Supplier and consumer *Standards and Quality* **128** pp 6-34
6. Zainetdinov R I 2018 Quality management 2 Seven Simple Statistical Quality Tools Moscow Russian Academy of Railways. Faculty of leadership training 36
7. Alekseeva S I 2002 The theory of nonlinear hereditary polymer media and its applications *Dis. for the degree of doct. tech. sciences* (Moscow)

IOP Conf. Series: Materials Science and Engineering 1023 (2021) 012006 doi:10.1088/1757-899X/1023/1/012006

Influence of laser steel quenching on the service life of friction units

V P Biryukov, A P Savin and A N Prince

Mechanical Engineering Research Institute of the Russian Academy of Sciences, 4 Maly Kharitonyevskiy Pereulok, Moscow, 101990, Russia

Laser-52@yandex.ru

Abstract. This paper presents the results of metallographic and tribological studies of 65Mn steel samples with laser and bulk quenching. A full factorial experiment was performed that allowed calculating the depth and width of the quenching zones with an error of no more than 3.28%. It is shown that laser hardening reduces the friction coefficients by 2-3 times and increases the jamming load compared to bulk steel quenching.

1. Introduction

Samples of 44MnSiVS6 steel in the form of disks with a thickness of 10 mm and a diameter of 110 mm with a surface roughness Ra 0.5 - 1μm were strengthened using a fiber laser system from IPG Photonics [1]. The maximum laser power was 3000 W, and the beam travel speed was 100 mm·s^{-1}. The treatment was performed with three types of laser spot. Round shape with a diameter of 4 mm, with a Gaussian distribution of radiation energy. A 4×4 mm rectangular spot with uniform energy distribution and a 4×4 mm rectangular spot with 16 radiation points, with a total power loss of 37.9% of the round spot. The depth, width, and microhardness of the steel hardening zones were 0.4×1.6 mm, 5800-8000 MPa, 0.3×1.4 mm, 4200-7300 MPa, 0.2×1.1 mm, and 3800-7200 MPa for round, rectangular, and spot spots, respectively. Treatment with a circular laser beam resulted in residual compression stresses of up to -506 MPa in the quenching zone to a depth of 400 μm.

Laser surface treatment was carried out on samples of powdered carbon steel containing 1% carbon PA - ZhGr with a porosity of 4%, 8% and 10%, as well as for comparison of U10 steel [2]. The powders were mixed in a mixer with an offset axis of rotation, pressed in a mold at a pressure of 400 MPa, then annealed in a vacuum furnace at a temperature of 900°C, re-compacted in a mold at a pressure of 500 or 550 MPa, and finally sintered in a vacuum at a temperature of 1150°C. Laser surface treatment was performed on samples in the form of parallelepipeds 25×25×200 mm in size using an industrial additive machine Optomec LENS 850-R with a 1 kW fiber laser in argon. The diameter of the spot of the laser beam during laser processing was assumed to be 1.5, 2, 2.5 and 4 mm. The laser radiation power was 251-502 W, and the beam travel speed was 8-12 mm·s^{-1}. In contrast to U10 steel, microstructure analysis in powder steels allowed us to divide the hardening region into 4 zones: I - melting with reduced porosity, II - melting with initial porosity, III - solidification, and IV-base. Structure in zones I and II, martensitic microhardness is high, up to 1000 $HV_{0.05}$. Porosity does not affect the microhardness of the surface layer, but it affects the depth of hardening, which is greater with lower porosity, and for a sample of U10 steel, the depth of hardening corresponds to the depth of the thermal impact zone.

For laser hardening, samples of AISI 1538MV steel with a diameter of 66 mm and a thickness of 5 mm with a hardness of 23.31 ± 1.75 HRC were used [3]. Laser heat treatment was performed on a Laserline LDF 10.00-100 diode laser with a maximum radiation power of 10 kW, with a three-coordinate and angular beam control system with a diameter of 3.5 mm, with a linear speed of 168 $mm \cdot s^{-1}$. The surface hardness of the samples was 51.5 ± 2.5 and 52.7 ± 1 HRC, respectively, after laser and bulk quenching. The depth of the laser hardening zone was 760 μm. Friction and wear tests were performed in accordance with ASTM G99 on a CSM tribometer, S/N 18-259 according to the disk-ball scheme (WC 5mm, 1500 HV). The following test parameters were used: load 5 N, sliding speed 0.10 $m \cdot s^{-1}$, friction path 100, 200, 300 and 500 m, track radii 4, 6, 8 and 10 mm. The sample of the original steel showed a constant average sliding friction coefficient of 0.37 at a distance of 200 m, and then this value increased significantly to 0.45 at a friction path of 500 m. The laser-quenched sample showed the lowest coefficient of friction of 0.295 for sliding distances of 100 and 200 m, and then the coefficient of friction increased to 0.361 for a friction path of 300 and 500 m. The depth of the wear hole was 540, 430, and 385 μm for the initial, volume-hardened, and laser-hardened steel, respectively.

2. Materials and equipment

Laser processing of 65Mn steel samples with dimensions of 12×20×70 mm was performed at the technological complex of the IMASH RAN collective center. The radiation power P=700-1000 W, the processing speed V=7-10 $mm \cdot s^{-1}$, and the beam diameter d=3.6-5.0 mm were selected as variable parameters. As an additional factor, we considered beam scanning with a fixed frequency f=226 Hz. A resonant type device with an elastic element on which a mirror is fixed was used. Metallographic studies of deposited coatings were performed on a PMT-3 microhardness meter at a load of 0.98 N, a metallographic microscope, and an AM425 digital microscope.

Tribological tests were carried out according to the scheme plane (heat – strengthened laser beam sample) - sleeve, steel 40Kh (49-52 HRC). The sliding speed and pressure on the sample varied discretely in the range of 0.5–3.5 $m \cdot s^{-1}$ and 1-5 MPa, respectively. TP22C oil was used as a lubricant.

The influence of processing modes on the parameters of hardened tracks was determined using the full factor experiment (FFE) method. To construct mathematical models, the depth H and width B of the laser hardening zones were considered as the system responses. Table 1 shows the levels of experimental factors.

Table1. Levels of experimental factors.

Factor z_i	Upper level of factor z_i^+	Lower level of factor z_i^-	Center of the plan z_i^0	Variation Interval λ_i	Dependence of the encoded variable on the natural
P (W)	1000	700	850	150	$x_1 = \frac{P_i - 850}{150}$
V (mm s^{-1})	10	7	8.5	1.5	$x_i = \frac{V_i - 8.5}{1.5}$
d (mm)	5.0	3.6	4.3	0.7	$x_i = \frac{d_i - 4.3}{0.7}$

Since FFE 2^3 was performed, the number of experiments was 8 for each series.
The regression equation has the form:

$$y = b_0 + b_1x_1 + b_2x_2 + b_3x_3 + b_{13}x_1x_3 + b_{13}x_1x_3 + b_{23}x_2x_3 + b_{123}x_1x_2x_3 \quad (1)$$

Where, y – system response; x_i – is the levels of factors; b – coefficients of the regression equation.

3. Results of experiments and calculations

Figure 1 shows the micro-sections of the laser thermal hardening zones of 65Mn steel obtained at the radiation power P = 1000 W, the beam travel speed V = 10 mm·s^{-1}, and the beam diameter d = 5.0 mm. With transverse beam vibrations with a frequency of 226 Hz and the same processing modes, the area of the quenching zone increases by 1.56 times (figure 1 (b)). Based on the results of metallography of all samples, regression equations for determining the parameters of laser hardening zones are obtained.

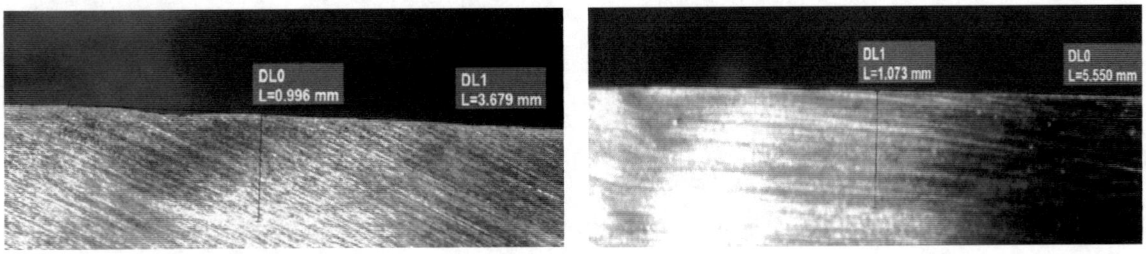

(a) (b)

Figure 1. Micro-Sections of laser thermal hardening zones of steel 65Mn: P = 1000 W, V = 10 mm·s^{-1}, d = 5.0 mm: a – defocused beam, b – scanning with a frequency of 226 Hz.

The equation for determining the depth of quenching without scanning, H has the form:

$$H = 0.917 + 0.08775x_1 - 0.0445x_2 + 0.09375x_3 - 0.02175x_1x_2 + 0.0325x_1x_3 - 0.02025x_2x_3 \quad (2)$$

Depth of hardening with transverse oscillations of the beam, Hs:

$$Hs = 0.866125 + 0.090875x_1 - 0.032125x_2 + 0.097875x_3 - 0.026875x_1x_2 + \\ +0.027625x_1x_3 - 0.031375x_2x_3 \quad (3)$$

Width of laser thermal hardening zones without scanning the beam, B:

$$B = 3.512625 + 0.122875x_1 - 0.088375x_2 + 0.330125x_3 - 0.041125x_1x_2 + \\ +0.079375x_1x_3 - 0.022875x_2x_3 \quad (4)$$

Width of quenching zones with transverse vibrations of the beam, Bs:

$$Bs = 5.97875 + 0.3265x_1 - 0.204x_2 + 0.23125x_3 - 0.02325x_1x_2 + 0.163x_1x_3 + \\ +0.066x_2x_3 \quad (5)$$

The regression equations (2-5) are used for calculations that are compared with the results of the experiment. The calculated values differ from the actual values of the depth and width of the quenching zones by no more than 3.28%.

The obtained regression models of H (P,V) and B (P,V) dependences are entered into the MsExcel table editor and comparative surfaces for these functions are constructed (figure 2) for defocusing a 5.0 mm laser spot. The radiation power has a predominant influence on the geometric parameters of the quenching zones. With increasing power, the width and depth of the quenching zone increase. As the speed of movement increases, the depth and width of the hardened zones decreases. As the laser defocusing increases, the depth and width of the hardened zones increases.

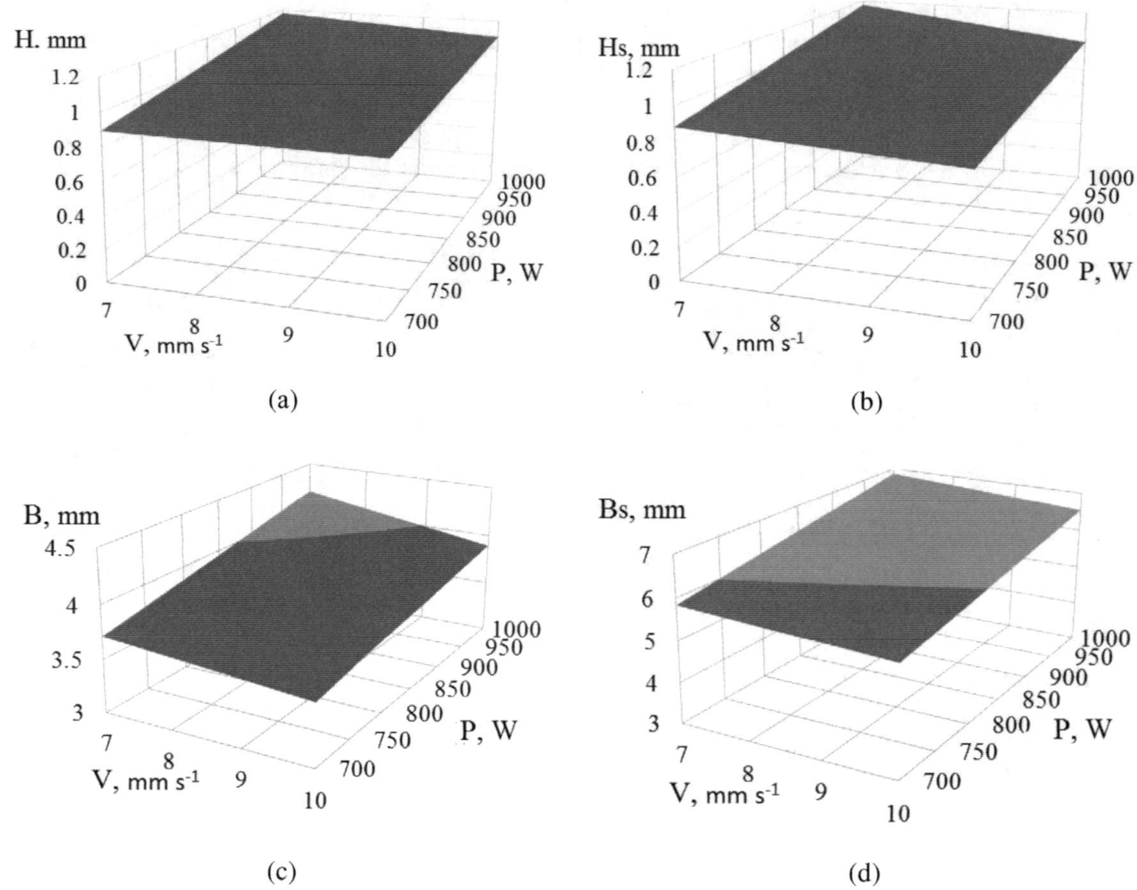

Figure 2. Dependence of the depth (a, b) and width (c, d) of the laser quenching zones of 65Mn steel on the processing speed and power at a beam diameter of 5.0 mm: (a, c) –with a defocused beam, (b, d) with a scanning beam.

The microhardness of the hardened zones varied in the range of 7780 – 9760 MPa depending on the treatment modes. With laser thermal hardening at high power and low beam speed, the microhardness of the quenching zone is reduced by 450-760 MPa. However, it significantly exceeds the microhardness of volume-hardened steel 65Mn by 1600-2800 MPa.

One of the most important characteristics of friction units is the low coefficient of friction, which affects the performance of fuel and lubricants in the operation of transport equipment. As a rule, the friction pair with a low coefficient of friction have a high load binding. The advantage of laser thermal hardening technologies in comparison with bulk quenching is shown in the absence or minimal residual deformations of parts after quenching. For example, when volumetric hardening of long parts, overhead sliding guides, residual deformations reach 3.5 mm per 1 m length. For further mechanical processing, straightening of the workpieces is required. The allowance for grinding such straightened workpieces is 0.7 - 0.8 mm per side. With laser thermal hardening, you can reduce the allowances to 0.1 mm per side and increase the productivity of subsequent machining.

Figure 3 shows the dependence of the friction coefficients on the pressure and sliding speed in the friction pair.

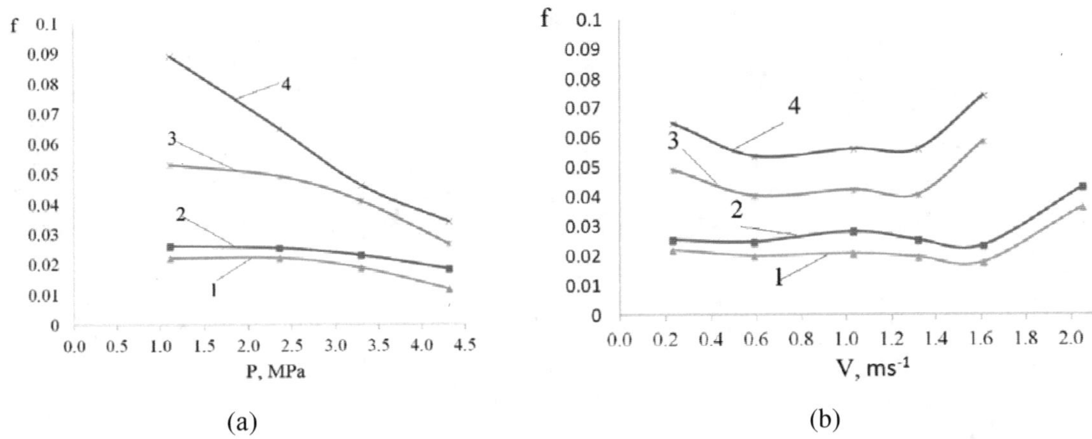

Figure 3. Dependence of friction coefficients on pressure (a) and sliding speed (b): 1 – LHT track pitch 5 mm, 2 = LHT track pitch 4 mm, 3- VHT, 52-57HRC, 4 – VHT, 48-51 HRC.

Curves 1 and 2 are obtained by friction of laser-quenched samples with transverse beam vibrations in steps of 5 and 4 mm, respectively. Reducing the step of superposition of laser thermal hardening zones (LHT) led to an increase in the area of the tempering zones between the tracks and, as a result, to an increase in the coefficient of friction. The value of the coefficient of friction of volume-hardened (VHT) samples was affected by the hardness, the higher it is, the lower the coefficient of friction. Regardless of the type of hardening with increasing pressure, the friction coefficients decreased in the studied load range. An increase in the sliding speed from the samples to 1.3 $m \cdot s^{-1}$ led to a sharp increase in the coefficient of friction. Samples with laser hardening had low friction coefficients of 0.02-0.03 and at a speed of 1.6 $m \cdot s^{-1}$.

Figure 4 shows the patterns of changes in the jamming load as a function of the sample sliding speed. The increase in sliding speed leads to a decrease in load binding for all types of hardening. Samples with laser thermal hardening withstood a load of 4.7 MPa before jamming at a sliding speed of 1.7 $m \cdot s^{-1}$, which is almost 2 times higher than volume-hardened samples. For parts operating under high contact pressures, the jamming load is an important criterion for their durability. No less important criterion is the wear rate of friction pairs, which determines the service life of components and mechanisms of machines and aggregates. The amount of wear intensity is affected by the mechanical properties of the friction surfaces, operating loads, sliding or rolling speeds, operating conditions and lubrication of mechanisms. Traditional technologies for hardening parts are energy-intensive and do not always meet the requirements of environmental cleanliness of the process. Modern laser systems are equipped with robotic systems or multi-coordinate systems with numerical control (CNC). Technological processes for laser modification of friction surfaces are environmentally friendly. In laser thermal hardening with a round spot, the time of beam exposure in the center of the spot is determined by the ratio of the beam diameter to the speed of its movement *d/V*, and at the edges of the spot tends to zero. In addition, the radiation power is distributed unevenly across the spot. There are one or more radiation maxima. To equalize the exposure time and power density across the spot, a cross – beam scan is applied, in which the area of the laser-hardened layer increases by 1.6-2.1 times, which means the process productivity. When quenching with a round beam, tempering zones with a width of 0.5-1.8 mm are formed along the edges of a single track. Oscillating beam hardening reduces these zones to a width of 0.1-0.18 mm. Therefore, the paper shows the effect of the quenching track step on the friction coefficients and the wear rate of samples. In table 2 shows experimental data on the wear rate of heat-strengthened laser beam samples with different steps and volume quenching of steel 65Mn to a hardness of 52-57HRC.

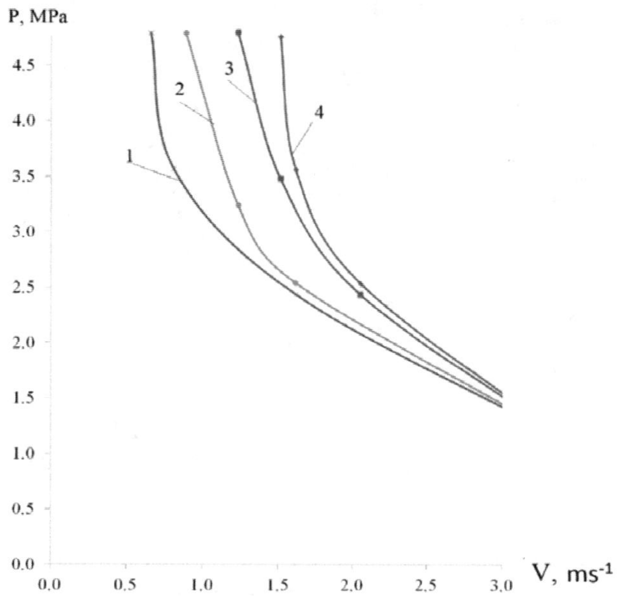

Figure 4. Dependence of the jamming load on the sliding speed: 1 - VHT, 48-51 HRC, 2 – VHT, 52-57HRC, 3 – LHT step of tracks 4 mm, 4 – LHT step of tracks 5 mm.

Table 2. Wear rate of steel 65Mn.

Type processing of steel 65Mn	Wear rate sample, I_1 and counter-sample, I_2		Wear rate sample I_{m1} and counter-sample I_{m2}	
	$I_1 \cdot 10^{-9}$	$I_2 \cdot 10^{-9}$	$I_{m1} \cdot 10^{-9}$	$I_{m2} \cdot 10^{-9}$
LHT, step - 4 mm	0.328 0.361 0.349	0.881 0.862 0.914	0.346	0.885
LHT, step - 5 mm	0.293 0.269 0.327	0.856 0.925 0.897	0.296	0.892
VHT 52-57 HRC	0.412 0.424 0.441	0.897 0.832 0.894	0.425	0.861

The wear rate of samples with laser hardening in steps of 4 and 5 mm is 22 and 43% lower than that of samples with bulk hardening. It should be noted that the wear rate of 40Kh hardened steel counter-tiles increases up to 3.6% in the friction pair with laser-hardened samples in 5 mm increments.

4. Conclusions

The technology of laser thermal strengthening of 65Mn steel with transverse beam vibrations has been developed, which allows to increase the productivity of the process by 1.6-2.1 times, depending on the laser processing modes. The friction coefficients of samples with laser quenching are 2-3 times lower, and the jamming load at a sliding speed of 1.7 $m \cdot s^{-1}$ is 2 times higher compared to bulk quenching. The wear rate of laser-quenched samples with different track spacing decreased by 22 and 43% compared to samples after bulk quenching.

References

[1] Dewi H S, Fischer A, Volpp J, Niendorf T and Kaplan F H 2020 *Optics and Laser Technology* **127**

[2] Oglezneva S, Morozov E and Usynin E 2019 *MATEC Web of Conf.* **298**
[3] Carrera-Espinoza R, Rojo Valerio A and Prado Villasana J *et al* 2020 *Advances in Materials Science and Engineering*

IOP Conf. Series: Materials Science and Engineering 1023 (2021) 012007 doi:10.1088/1757-899X/1023/1/012007

Determination of properties of cermet coatings obtained by additive laser technologies

V P Biryukov

Mechanical Engineering Research Institute of the Russian Academy of Sciences, 4 Maly Kharitonyevskiy Pereulok, Moscow, 101990, Russia

Laser-52@yandex.ru

Abstract. The paper presents the results of metallographic and tribological studies of 40Kh steel and multi component coating (Ni-Cr-B-Si, Co-Cr-W-Ni-B-Si) with the addition of 5 and 10 vol.% nano- tungsten carbide powder to the charge. The regularities of changes in the friction coefficients from pressure, sliding speed, jamming load from sliding speed and wear intensity of steel and surfaced coatings are shown.

1. Introduction

Additive laser processes play an important role in modern machine-building production. Surfacing experiments were performed using a 4 kW IPG fiber laser. The optical head with a coaxial nozzle was moved by a 6-axis robotic system [1]. Low-carbon steel samples with dimensions of 100×45×45 mm were used as a substrate. For laser surfacing, a composite powder material containing spherical particles of H13 steel with a size of 50 – 150 μm, and crushed titanium carbide (TiC) 30, 40, 50, 60, 70 and 80% of the charge composition with a size of 60 – 150 mm was used. The total feed rate of the powder was 9-10 $g \cdot min^{-1}$. Typical technological parameters used for surfacing were the laser power density of 180-220 $W \cdot mm^{-2}$, the spot diameter of 4 mm, and the travel speed of 0.2-0.3 $m \cdot min^{-1}$. In addition, the protective gas argon was used, at a flow rate of 10 $l \cdot min^{-1}$. The first layer was deposited with H13 powder. The highest average hardness of the TIC/H13 composite is 1365 HV at 80% TiC, which is more than twice the hardness of H13 - 644 HV substrate. As the TiC content increases, the average hardness increases. Thus, at 30% TiC in the charge, the average hardness of the deposited layers is 787 HV.

For laser surfacing of WC-10Co-4Cr powder with a particle size of 11-45 μm, flat samples of AISI–SAE 1020 steel with dimensions of 76×25×6 mm were used [2]. Processing was performed using a 4kW Yb: YAG Trumpf TruDisk 6002 laser system equipped with a powder feeder, a beam with a diameter of 3.8 mm with a pulse duration of 0.014 s and a frequency of 50 Hz. The energy density of the laser beam was 24.32 and 39 $j \cdot mm^{-2}$ at a speed of 10 $mm \cdot s^{-1}$. The results of metallography showed that at an energy value of 24 $j \cdot mm^{-2}$, the porosity was about 11%, while the mixing of the coating with the base was at the level of 15%. Increasing the energy density reduces porosity by up to 4% and increases mixing by up to 25%. The maximum microhardness value of 1200 HV was obtained at an energy density of 24 $j \cdot mm^{-2}$. As the energy density increases, the grain size increases and the microhardness drops to 1000 HV.

Samples made of SS 304 steel with dimensions of 50×25×6 mm were surfaced using Inconel 718 and TiC powders with particle sizes of 43-100 μm and 5-20 μm, respectively [3]. The first batch of pre-coated samples using a 1 mm thick binder was deposited at a radiation power of 1200 W, a travel speed of 200-1200 mm·min^{-1}, and a laser spot diameter of 3 mm. The content of titanium carbide in the charge for surfacing was 30, 50 and 70%. The second batch of samples was processed using a powder feeder, according to the following modes: powder consumption 8-28 g·min^{-1}, laser beam diameter 2.2 mm, radiation power 400-1200 W. It was found that the lifetime of the molten bath in the range of $0.25 < \tau \leq 0.45$ s leads to a good quality of the deposited rollers. Longer life time of the melt of 0.45 with the above, leads to complete decomposition of TiC particles.

Stainless steel 304 plate with dimensions of 160 × 80 × 10 mm is used as substrates for laser surfacing of sintered WC-12Co composite powder with a spherical particle size of 125-180 μm [4]. The treatment was performed on a laser system (LMD8060) containing a LaserLine LDF 3.000-60 diode laser. The following modes were used: radiation power 900, 1100, 1300, 1500, 1700 and 1900 W, beam travel speed 500 mm·min^{-1}, laser spot diameter 2 mm, powder consumption 0.5g·min^{-1}, track overlap coefficient 50%, thickness of the deposited layer for three passes 2 mm. Tribological tests of coatings at room temperature were performed according to the scheme disk (deposited sample) – ball (Al_2O_3 with a diameter of 6 mm), on a Tribometer (TNT-1000), under friction conditions without lubricant. The sliding speed was 0.2 m·s^{-1} and the friction path was 200 m. Each sample was tested for 20 minutes under a load of 5 N. The Microhardness of the first deposited layer increased from the substrate, and reached 1000-1200 $HV_{0.3}$ at its surface, and was approximately 6 times higher than the average microhardness of the base material. The microhardness of the second and third layers also increased from 1200-1400 to 1500-1600 $HV_{0.3}$. The Coating obtained at a laser power of 1500 W had the lowest coefficient of friction of 0.55 and the wear rate (2.15±0.31) ×10^{-7} mm^3·(N·m)$^{-1}$.

The aim of this work is to determine the tribological characteristics of multi component coatings with the addition of nano-powder of tungsten carbide.

2. Materials and equipment

The laser system of the IMASH RAN collective center was used in experimental studies. The samples were made of 40Kh steel with dimensions of 15×20×70 mm. Powders based on Nickel PR-NiKh13SiB3 and cobalt PR-CoKh30Ni6WSiB in a ratio of 2:1 with a particle size of 40-150 μm, respectively, were selected for the production of the charge. Nano tungsten carbide powder WC with a particle size of 40-100 nm. The composition of the powders can be represented as (Ni-Cr-B-Si, Co-Cr-W-Ni-B-Si) + WC 5 and 10 vol.%. Slip coatings were applied with a thickness of 0.85-1.0 mm. An aqueous solution of hydrooxyethylcellulose was used as the binding material. The radiation power P = 800-1000 W, the processing speed V= 5-10 mm·s^{-1}, and the beam diameter d = 2,5-3,5 mm were chosen as the variable parameters. As an additional factor, we considered scanning a beam with a fixed frequency f=225 Hz. Metallographic studies of deposited coatings were performed on a PMT-3 microhardness meter at a load of 0.98 N, an Altami MET 1C metallographic microscope, and an AM413ML digital microscope. The structure and chemical composition of the deposited layers were studied using a Tescan VEGA 3 SBH scanning electron microscope with an energy-dispersive analysis system using reflected and secondary electron modes. To determine the tribological characteristics of the deposited samples, a test was performed at normal temperature according to the plane (deposited sample) – ring scheme (50KhVA steel, 49-52). The sliding speed and pressure on the sample varied discretely in the range of 0.25-3.0 m·s^{-1} and 1-5 MPa, respectively. TP22C oil was used as a lubricant.

The radiation power P=700-1000 W, the processing speed V=7-10 mm·s^{-1}, and the beam diameter d=2.5-3.5 mm were chosen as the variable parameters. As an additional factor, we considered beam scanning with a fixed frequency f=225 Hz. To construct mathematical models, the height H and width B of the deposited rollers were considered as the system responses when performing a full factorial experiment (FFE). Since the range of variable parameters has a rather narrow range, a linear dependence of changes in the experimental factors is assumed. Table 1 shows the levels of experimental factors. Since FFE 2^3 was performed, the number of experiments was 8 for each series.

The regression equation has the form:

$$y = b_0 + b_1x_1 + b_2x_2 + b_3x_3 + b_{13}x_1x_3 + b_{13}x_1x_3 + b_{23}x_2x_3 + b_{123}x_1x_2x_3 \quad (1)$$

Where y – system response; x_i –is the levels of factors; b – coefficients of the regression equation.

Table1. Levels of experiment factors.

Factor z_i	Upper level of factor z_i^+	Lower level of factor z_i^-	Center of the plan z_i^0	Variation Interval λ_i	Dependence of the encoded variable on the natural
P (W)	1000	700	850	150	$x_1 = \frac{P_i - 850}{150}$
V (mm·s^{-1})	9	7	8	1	$x_i = V_i - 8$
d (mm)	3.5	2.5	3.0	0.5	$x_i = \frac{d_i - 3}{0.5}$

3. Results of experiments

Laser surfacing of samples was performed under optimal conditions with a defocused beam and with transverse beam vibrations normal to the laser processing speed vector. Figure 1 (a, b) shows micro-sections of deposited tracks with a 10 vol. % content of nano-tungsten carbide with dimensions of 0.87×2.34 mm, hardness (980-1040 HV), and 0.83×3.56 mm - (1030-1120 HV), obtained by an unfocused beam and a scanning beam with a frequency of 225 Hz, respectively. The penetration zone of the substrate during processing with a defocused beam and a scanning beam was 189 and 86 μm, respectively. The cross sectional area of a single deposited layer when scanning the beam is 1.57 times larger than when surfacing with an unfocused beam.

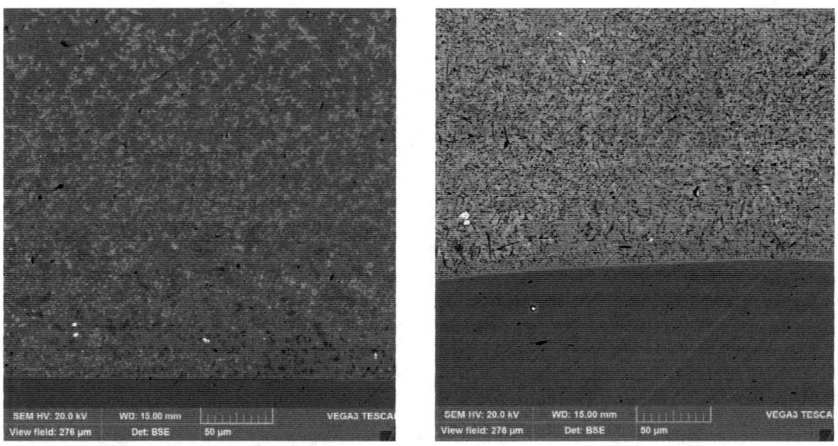

(a) (b)

Figure 1. Microstructures of deposited coatings containing 10 vol. % nano-tungsten carbide: a-defocused beam surfacing; b-scanning beam surfacing.

The equation for determining the height of the roller without scanning, H has the form:

$$H = 0.925625 + 0.156875x_1 - 0.119375x_2 + 0.115625x_3 - 0.048125x_1x_2 + \\ +0.031875x_1x_3 - 0.054375x_2x_3 \quad (2)$$

The height of the roller when welding with transverse oscillation of the beam, Hs:

$$Hs = 0.93375 + 0.09625x_1 - 0.02625x_2 + 0.02375x_3 + 0.01625x_1x_3 - 0.07625x_2x_3 - \\ -0.04875x_1x_2x_3 \quad (3)$$

Width of the deposited roller without scanning the beam, *B*:

$$B = 1.7125 + 0.235x_1 - 0.15x_2 + 0.0475x_3 - 0.1075x_1x_2 - 0.045x_1x_3 - 0.065x_2x_3 \quad (4)$$

Width of the deposited rollers with transverse vibrations of the *Bs* beam:

$$Bs = 3.61625 + 0.35875x_1 - 0.37875x_2 + 0.07125x_3 - 0.03125x_1x_2 + \\ +0.21375x_1x_3 + 0.15625x_2x_3 \quad (5)$$

The regression equations (2-5) are used for calculations that are compared with the results of the experiment. The calculated values differ from the actual values of the depth and width of the quenching zones by no more than 3.14%.

For dependencies of type *H (P, V), B (P, V)*, comparative surfaces are constructed using the MsExcel program (figure 2) with a spot diameter of 2.5 mm.

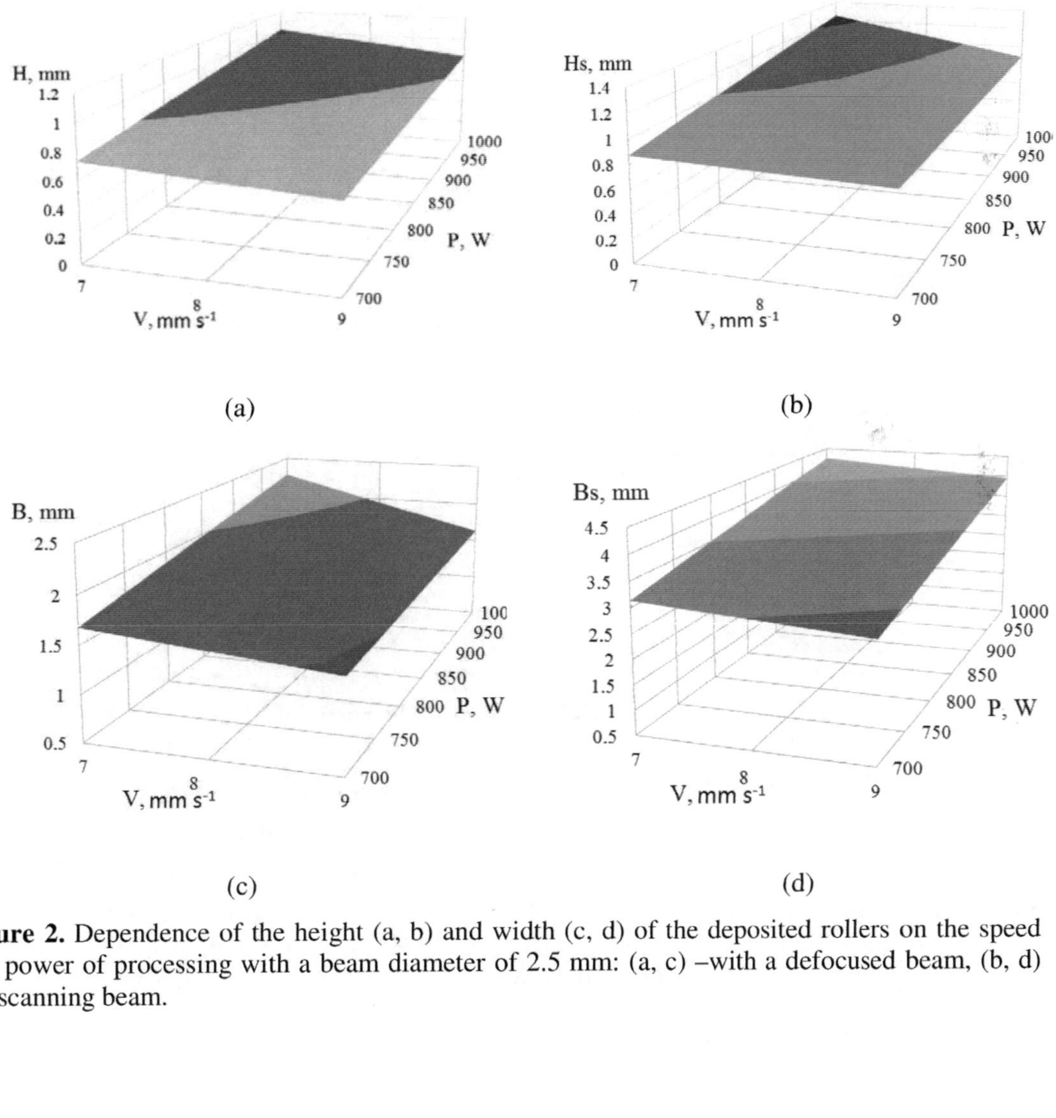

Figure 2. Dependence of the height (a, b) and width (c, d) of the deposited rollers on the speed and power of processing with a beam diameter of 2.5 mm: (a, c) –with a defocused beam, (b, d) are scanning beam.

The radiation power has a predominant influence on the geometric parameters of the deposited rollers. With increasing power, the width and height of the surfaced tracks grow. As the speed of movement increases, the depth and width of the rollers decreases. With increasing defocusing of laser radiation, the depth and width of the rollers increases.

One of the most important characteristics of friction units is the low coefficient of friction, which affects the performance of fuel and lubricants in the operation of transport equipment. As a rule, the friction pair with a low coefficient of friction has a high load binding.

One of the most important characteristics of friction units is the low coefficient of friction, which affects the performance of fuel and lubricants in the operation of transport equipment. As a rule, the friction pair with a low coefficient of friction has a high load binding. The dependence of the friction coefficients of 40Kh steel in the improved and hardened state and the deposited coatings is shown in figure 3.

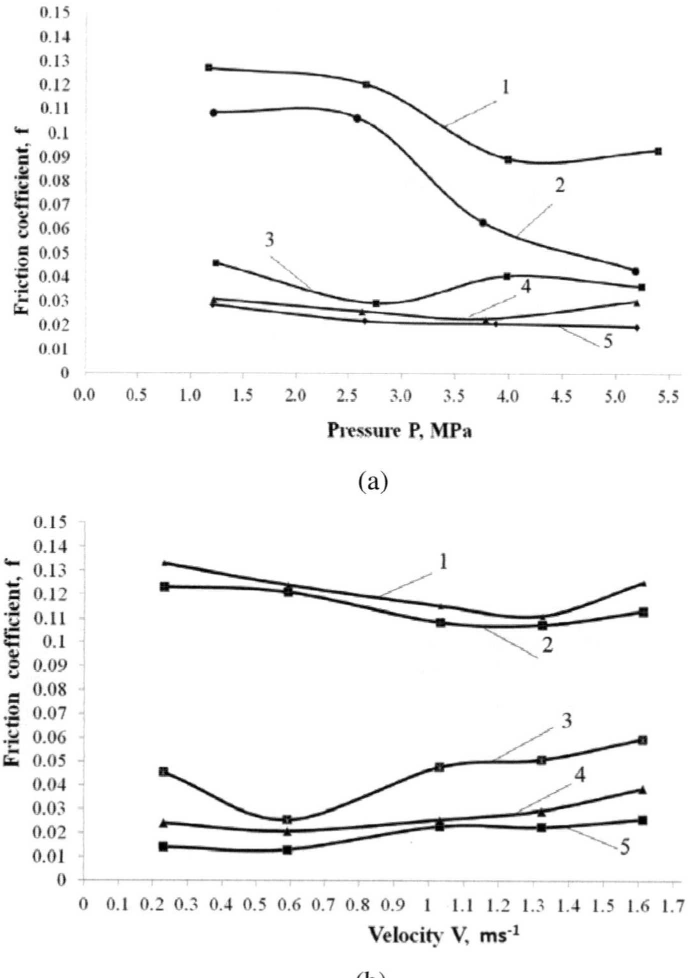

Figure 3. Dependence of friction coefficients on pressure (a) and sliding speed (b): 1 – 40Kh steel (250-296HV), 2 – 40Kh steel (450-510HV), 3 – Ni-Cr-B-Si, Co-Cr-W-Ni-B-Si (750-860HV), 4 – (Ni-Cr-B-Si, Co-Cr-W-Ni-B-Si) + 5 WC vol.% (850-970HV), (Ni-CR-B-Si, co-CR-W-NI-b-SI) + 10 WC vol.% (890-1270HV).

As the load increases from 1.2 to 4.0 MPa on improved and hardened 40Kh steel samples, the coefficient of friction drops from 0.13 and 0.11 to 0.1 and 0.07, respectively. With a further increase in the load for the improved sample, the coefficient of friction increases, and for the hardened one it continues to decrease to 0.048. The coefficient of friction for the multi component coating varies within 0.03-0.043. The Minimum coefficient of friction 0.01-0.02 was obtained on a coating with additives of 10 vol% WC nano powder. With an increase in the sliding speed from 0.25 to 1.35 $m \cdot s^{-1}$ under constant load, the coefficient of friction for 40Kh steel decreases from 0.12-0.13 to 0.1-0.11. With a further increase in speed to 1.6 $m \cdot s^{-1}$, it increases slightly. For deposited coatings in the range of 0.6-1.6 $m \cdot s^{-1}$, the coefficient of friction increases smoothly.

Figure 4 shows the patterns of changes in the jamming load from the sliding speed. Improved and hardened 40Kh steel samples are inferior to surfaced multicomponent coatings and with the addition of nano-tungsten carbide powder. At a pressure of 5.5 MPa, jamming occurs at a speed of 3-5 times and 0.25-2 times lower for an improved and hardened 40Kh steel sample, compared to surfacing with a multicomponent coating and with additives of 10 W vol.% respectively.

For figure 5 wear rates of 40Kh steel samples with different heat treatment and deposited coatings are shown. Wear resistance, the inverse of the wear intensity, increases for many component coatings by 30 and 25% and 57 and 62% compared to improved and hardened 40Kh steel with an addition of 5 and 10 vol.% to charge nano powder WC respectively.

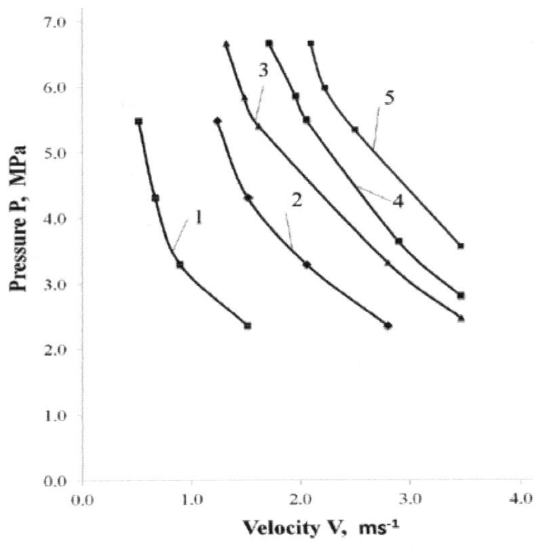

Figure 4. Dependence of the jamming load on the sliding speed: 1 – 40Kh steel (250-296HV), 2 – 40Kh steel (450-510HV), 3 - Ni-Cr-B-Si, Co-Cr-W-Ni-B-Si, 4- (Ni-Cr-B-Si, Co-Cr-W-Ni-B-Si) + 5 WC vol.% (850-970HV), 5 - (Ni-Cr-B-Si, co-CR-W-Ni-B-Si) + 10 WC VOL.% (890-1270HV).

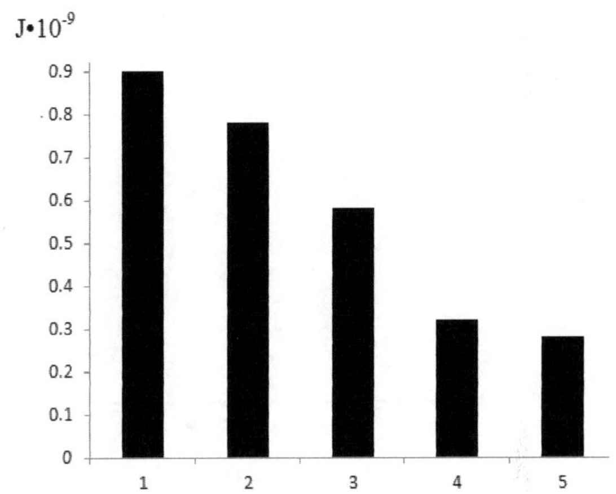

Figure 5. wear rates of samples: 1 – 40Kh steel (250-296HV), 2 – 40Kh steel (450-510HV), 3 - Ni-Cr-B-Si, Co-Cr-W-Ni-B-Si 4- (Ni-Cr-B-Si, Co-Cr-W-Ni-B-Si) + 5 WC vol.% (850-970HV), 5 - (Ni-Cr-B-Si, Co-CR-W-Ni-B-Si) + 10 WC VOL.% (890-1270HV).

4. Conclusions

The technology of laser surfacing of multi component coatings with additives of 5 and 10 vol.% nano-tungsten carbide has been develop decreases the wear resistance of improved and hardened 40Kh steel by 25-62% and the load of jamming up to 5 times at a pressure of 5.5 MPa. Coatings with nano - tungsten carbides have the lowest friction coefficients of 0.01-0.02.

References

[1] Chen H, Lu Y, Sun Y, Wei Y, Wang X and Liu D 2020 *J. Surface and Coatings Technology* **395** pp 125867
[2] López-Baltazar E A, Ruiz-Luna H, Baltazar-Hernández V H, Ruiz-Mondragón J J, Ibarra-Medina J and Alvarado-Orozco J M 2019 *Metals* **9** p 1245
[3] Muvvala G, Mullick S and Nath A K 2020 *J. Surface and Coatings Technology* **399** pp 126100
[4] Miao H U, Tang J, Chen X, Nan Y E, Zhao X and Xu M 2020 *Transaction Nonferrous Metals Society of China* **30** p 1017–30

Determination of tribological characteristics of polymer materials

V P Biryukov and A N Prince

Mechanical Engineering Research Institute of the Russian Academy of Sciences, 4 Maly Kharitonyevskiy Pereulok, Moscow, 101990, Russia

Laser-52@yandex.ru

Abstract. The paper presents the results of tribological tests of thermoplastic polyurethanes (TPU) and rubber based on nitrile butadiene rubber (BNR) on rigidly fixed silicon carbide and free abrasive. The influence of pressure and sliding speed on the coefficients of friction is shown. For Elastollan 1164, the coefficient of friction and mass loss were 0.2-0.3 and 0.028 g, which is 5-7 times lower than for BNR.

1. Introduction

TPU samples (Ellastolan ET870-11V, BASF, Germany) were obtained by injection molding a sheet with a thickness of 4 mm [1]. In addition, samples of BR butadiene rubber and SBR styrene-butadiene rubber were used. The sheets were cut into blocks with dimensions of 5×5×3 mm. During the test, the sample slid along the abrasive paper unidirectionally at a length of 0.07 m. The sample was then lifted up and returned to its original position. This process was repeated 100 times. The friction path of the sample was 7.0 m. Sheets of silicon carbide SiC abrasive paper with different grain sizes 3000, 1000, 500, 240, and 80 with roughness Ra 1.94, 4.02, 8.64, 17.8, and 45.7 μm, respectively, were glued to the fixed base. For three repeated measurements of the friction coefficient f of the TPU samples, there was no noticeable effect of the grain size of the abrasive paper ($1.05 \leq f \leq 1.12$). In contrast, the coefficient of friction for BC samples decreased from 0.96 to 0.69 with an increase in Ra at Ra $\leq$ 8.64 μm, and then it increased from 0.69 to 0.96 at Ra $\geq$ 8.64 μm. For SBC samples, the coefficient of friction decreased slightly from 0.91 to 0.84 with increasing Ra at Ra $\leq$ 17.8 μm and then increased from 0.84 to 0.98 at Ra$\geq$17.8 μm. The values of the friction coefficient for TPU exceeded 1.0 in comparison with the samples of BC and SBK, regardless of the Ra of the abrasive paper. The specific wear rate for all materials generally increased with increasing Ra. The wear rate for TPU was higher than for BC, but samples made of thermoplastic polyurethane had a higher wear resistance compared to SBC samples.

For testing, a thermoplastic polyurethane TPU (UT-74D) from Bayer, Germany, which has good wear resistance, mechanical strength and is widely used for sealing, in the automotive and footwear industry [2]. The height and diameter of the samples were 10 ± 0.05 mm and 30 ± 0.05 mm, respectively. Copper balls (H65, with Ra $\leq$ 0.2 μm, 10 ±0.03 mm in diameter) were used as a counter-sample. The friction pair simulated the actual operating conditions of the aft bearings lubricated by sea water. Tribological tests were performed according to the disk - ball scheme and using a multifunctional Tribometer (MFT5000, USA). The applied loads were 2, 6, 10 and 14 N, providing an initial average contact pressure of 14.34, 20.68, 24.52, 27.44 MPa in Hertz. The speed of rotation of the disk was 70 min^{-1} (0.04 $m \cdot s^{-1}$). the Ball glided along a circle with a radius of 6 mm. The TPU samples were irradiated with 60 Co radiation source, EL PONT Inc. China, with increasing absorbed radiation energy for each batch of 50, 150 and 250 kGf. Unirradiated samples were designated 0K. The coefficient of friction of TPU samples under various loads and the degree of irradiation of 0K and 50K TPU remained relatively stable, compared with samples 150K and 250K, at which it increased to 0.17-0.35 and 0.35 and 0.42, respectively, with increasing pressure.

To determine the tribological characteristics, mixtures of polypropylene and thermoplastic polyurethane (PP/TPU) were used in various weight ratios (25/75 and 75/25) with the addition of maleic anhydride PP-g-MA as a binder grafted to polypropylene. in quantity 0, 3, 5, 7, 9, 11 wt.h. per 100 wt.h. PP/TPU [3]. Friction and wear tests were performed according to the US standard ASTM G99 according to the scheme disk (a sample of a composite with a diameter of 29 mm and a thickness of 8 mm) - ball (Al_2O_3 with a diameter of 5 mm). The test load was 2.5 N, the sliding speed was 0.576 $m \cdot s^{-1}$ (1100 min^{-1}), the friction path was 345 m, and the duration was 10 min. Samples of PP25/TPU75 composites had a coefficient of friction of 0.39 - 0.6 and a wear RATE 1200 – 1800 × 10^{-6} $mm^3 \cdot N^{-1} \cdot m^{-1}$ with their maximum values at 5 wt.h. PP-g-MA. For the PP75/TPU25 mixture, the coefficient of friction was 0.39 - 0.4 and the wear rate 1000 - 1600 × 10^{-6} $mm^3 \cdot N^{-1} \cdot m^{-1}$ according to the results of wear tests, it was found that the PP75/TPU25 mixture with a content of 11 wt.h. PP-g-MA is superior to other mixtures.

2. Research materials and methods

For testing for friction and wear, thermoplastic polyurethanes (TPU) were selected, the technical characteristics of which are shown in table 1 for comparison with the polyurethanes of the selected rubber based on butadiene-nitrile rubber (BNR). Each of the 7 materials studied was a plate with dimensions of 20×70×2 mm, which was glued to plywood with dimensions of 20×70×12 mm.

Table 1. Technical characteristics of TPU.

Material brand	Hardness, Shore	Density, $g \cdot cm^{-3}$	Ultimate strength, MPa	Elongation at break, %
Elastollan 560 A	60 D	1.22	24	850
KOPEL KP 3340	40 D	1.15	24.5	850
Elastollan 1164 D 11	69 A	1.18	50	350
Elastollan 1154 D 10	53 A	1.17	50	450
Elastollan 1175 A 10 W	75 A	1.14	40	700
Elastollan 1185 A 10 W	83 A	1.16	40	700
BNR	75 A	1.3	25	425

A steel ring mandrel with sandpaper made of silicon carbide with a grain size of 120 μm pasted on its end was used as a counter-tile. Tests for friction and wear were performed according to the "plane (test sample) - ring" scheme. The sliding speed and pressure on the sample varied discretely in the range of 0.1–0.3 $m \cdot s^{-1}$ and 0.1–0.5 MPa, respectively. Wear tests with a free abrasive were performed when a flat sample was rubbed against the forming surface of a rubber disk. Quartz sand with a particle size of 0.2-0.6 mm was fed into the friction zone. The tests were performed at normal atmospheric pressure and temperature. The test load was 15 N, and the time was 5 minutes. Three samples of each material were tested. The amount of wear was determined as the average arithmetic weight loss for three samples.

3. Results of experimental studies

Figure 1 (a, b) shows the results in the form of dependences of the coefficient of friction on the normal pressure and sliding speed when tested on a fixed abrasive.

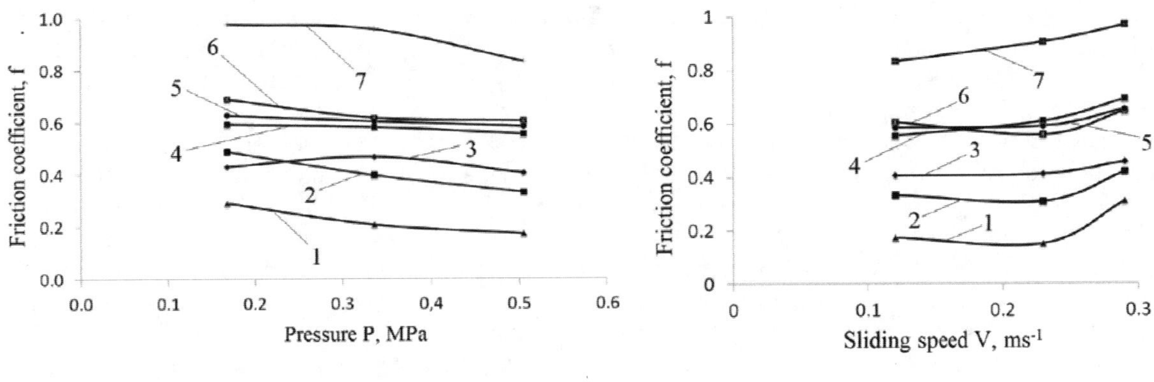

Figure 1. Dependence of polymer friction coefficients on pressure (a) and sliding speed (b): 1 – 1164, 2 – 1154, 3 – 560, 4 – 3340, 5 – 1185A, 6 – 1175A, 7 – BNR.

As the pressure increased, the values of the friction coefficients decreased for TPU1164, 1154, 1175A and BNR over the entire pressure range. The values of the friction coefficients of 0.6 and 0.61 remained constant for the TPU 3340 and 1185A, independent of pressure. For polymer 560, the coefficient of friction increased from 0.4 to 0.43 at a pressure of 0.34 MPa, and then decreased to 0.38 at a pressure of 0.5 MPa. An increase in the sliding speed from 0.12 to 0.24 $m \cdot s^{-1}$ led to a slight decrease in the friction coefficients for TPU1164, 1154, 1175A. With a further increase in the speed to 0.3 $m \cdot s^{-1}$, the coefficient of friction increased for all types of polymer materials. The BNR - 0.8 - 0.9 coefficient of friction increased proportionally over the entire range of sliding speeds.

Figure 2 shows the mass loss of samples from pressure.

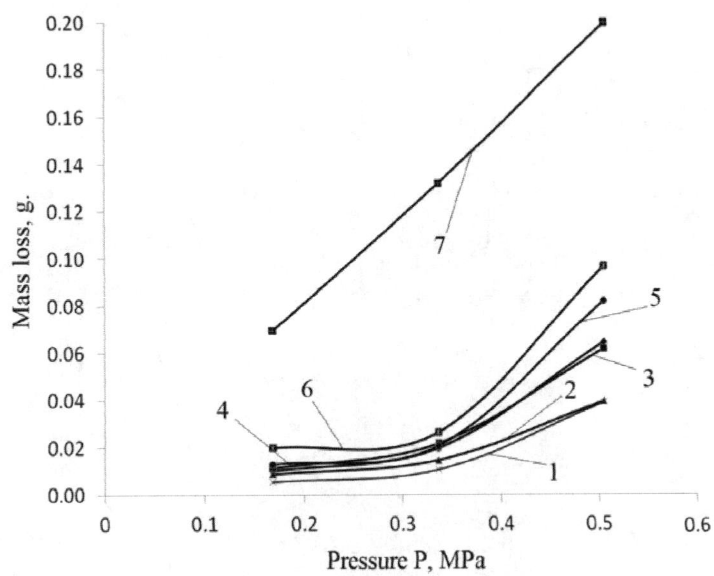

Figure 2. Dependence of the mass loss of the TPU sample on the pressure when testing on a fixed abrasive: 1 – 1164, 2 – 1154, 3 – 3340, 4 – 560, 5 – 1185A, 6 -1175A, 7 – BNR.

For the BNR, the mass loss increased in proportion to the pressure. The wear of TPU samples at low pressures up to 0.3 MPa is insignificant, with a further increase in pressure up to 0.52 MPa for Elastollan 1175 A and 1185 A increases to 0.078 and 0.087 g, respectively. The minimum mass loss was obtained for TPU 1164 0.004 g. At a pressure of 0.52 MPa, the minimum wear of 0.037 g was recorded in the Elastollan 1164 and 1154 samples. Figure 3 shows the morphology of the friction surfaces of the samples. The friction surface of the TPU560 (figure 3 (a)) has swells and rough delamination of the material, apparently as a result of repeated elastic and plastic deformation and fatigue separation of wear particles. On the surface of samples 3340, 1175A, 1185A (figure 3 (b), (d), (e)), risks are visible in the direction of sliding of the fixed abrasive grain. The undulation of the surface of the TPU 1164 and 1154 samples (figure 3 (c), (d)) indicates a gradual accumulation of damage and separation of wear particles by the fatigue mechanism. The friction surfaces of the BNR samples (figure 3 (g)) throughout had risks in the direction of sliding. It can be assumed that the prevailing mode of wear was micro-cutting with friction on the fixed abrasive grain. Figure 4 shows the profiles of wear holes of polymer materials and BNR rubber when tested with loose abrasive. Similar test results with minimal weight loss were shown by samples of TPU 1164 and 1154 with free and fixed abrasive. Samples of polymers 3340 and 560 when rubbed with free abrasive grain showed worse results than BNR. Samples 1185A and 1175A retained less weight loss compared to rubber.

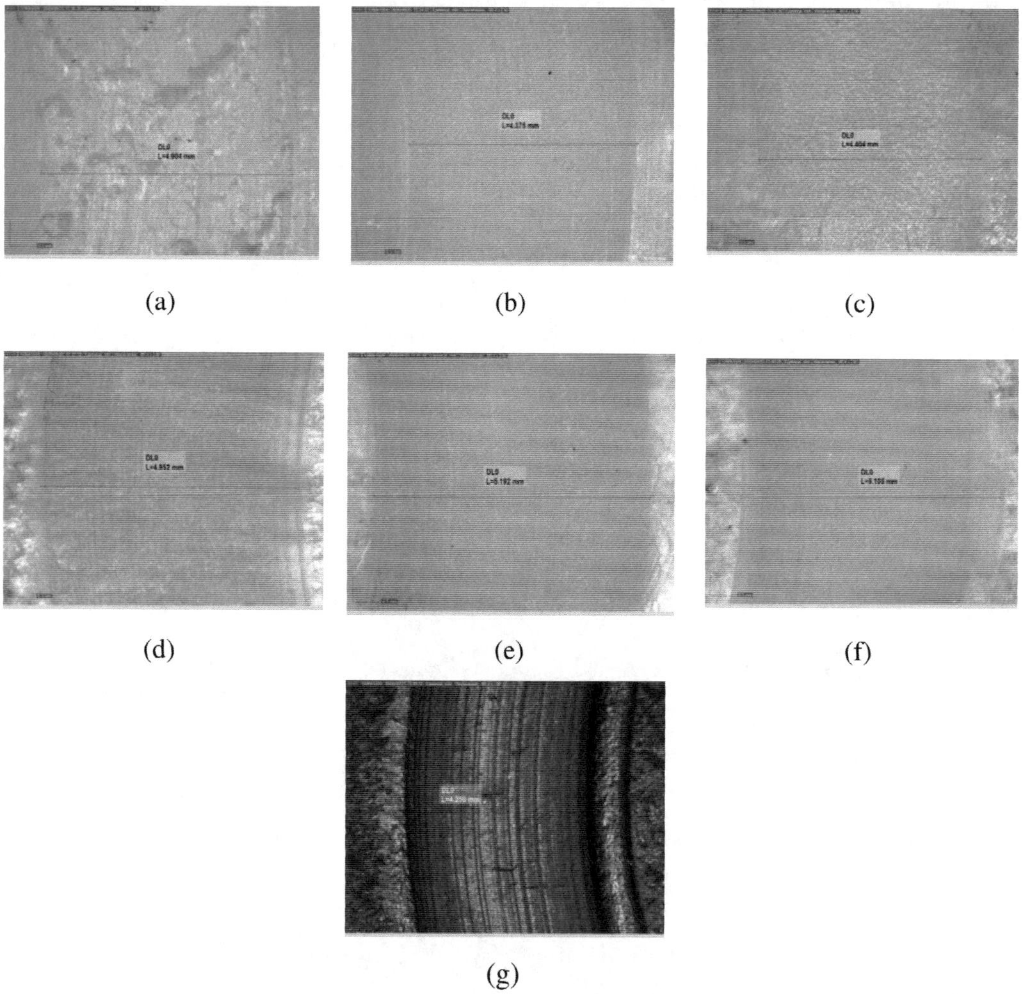

Figure 3. Morphology of the friction surface of TPU and BNR when tested with a fixed abrasive: (a) - 560, (b) – 3340, (c) - 1164, (d) - 1154, (e) – 1175A, (f) – 1185A, (g) – BNR.

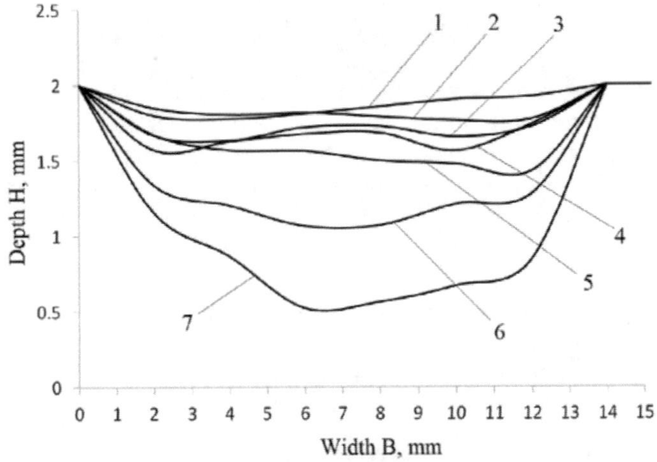

Figure 4. Hole profiles export of materials: 1 – 1164, 2 – 1154, 3 – 1185A, 4 – 1175A, 5 – BNR, 6 – 3340, 7 – 560.

Figure 5 shows fragments of wear holes of materials during friction on a free abrasive.

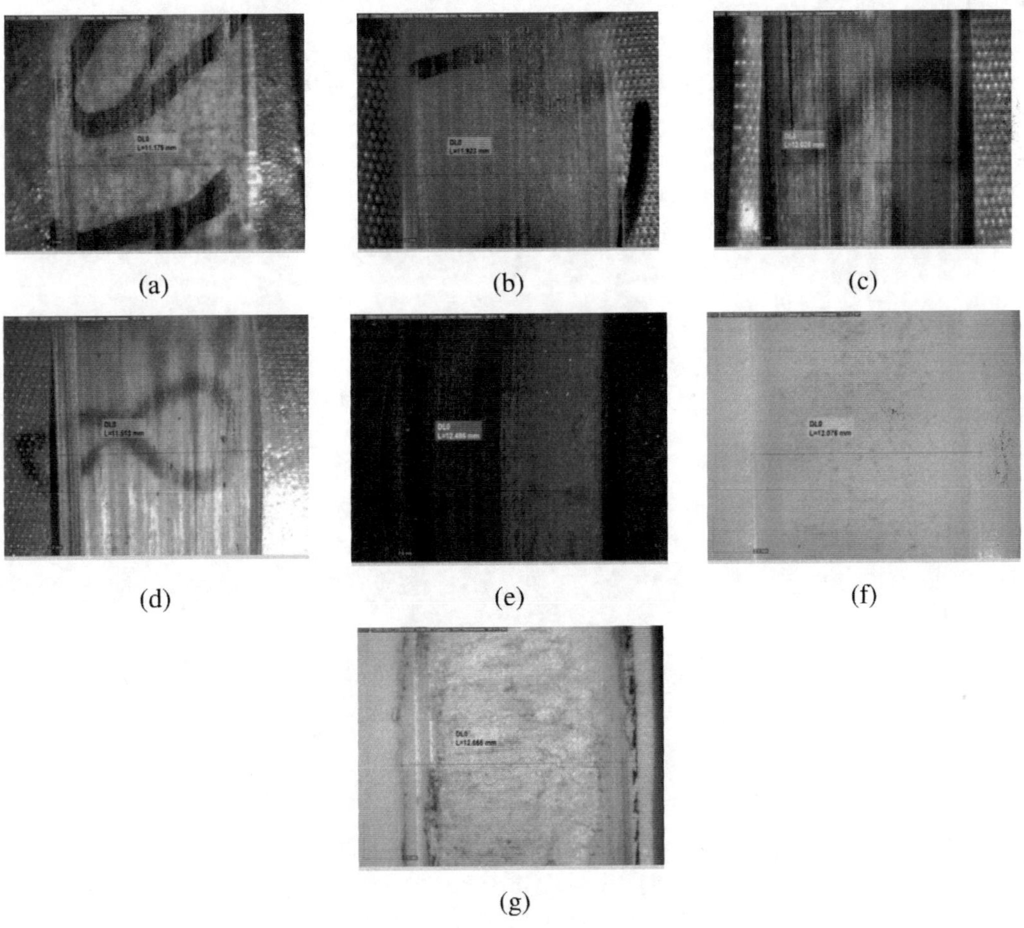

Figure 5. Morphology of the reunion of polymers and BNR in the free abrasive confession: (a) – 1164, (b) – 1154, (c)-1185A, (d) – 1175A, (e) – BNR, (f) – 3340, (g) – 560.

The friction surfaces of TPU 1164, 1154, 1185A, 1175A and BNR had longitudinal risks in the direction of sliding of the abrasive grain and the wear mechanism can be attributed to micro-cutting. The undulation of the surface of samples 3340, 560 indicates a gradual accumulation of damage and separation of wear particles by the fatigue mechanism. These materials are significantly inferior to the wear resistance of BNR when friction is not fixed with an abrasive, and can not be used to replace rubber products. Despite the different results, both test methods complemented each other and gave a complete description of polymer materials used in natural conditions during the operation of machines and mechanisms, when it is possible to affect the friction surface with fixed and free abrasive grain.

4. Conclusions

The regularities of changes in the coefficients of friction and mass loss of samples from pressure during friction on the fixed abrasive are obtained. Coefficient of friction for TPU 1164 – 0.2-0.3 and for BNR 0.8-1.0. The minimum weight loss was shown by TPU 1164 and 1154 when testing for wear with a free and fixed abrasive. Samples 3340, 560 had a greater mass loss than BNR when tested with free abrasive and cannot be used to replace it.

References

[1] Sato S, Yamaguchi T, Shibata K, Nishi T, Moriyasu K, Harano K and Hokkirigawa K 2020 *Biotribology* **23** p 100130
[2] Jiang S, Yuan C, Guo Z and Bai X 2019 *Tribology International* **136** pp 276–84
[3] Savaş S and Al-Obaidi AY 2018 *Tribology Transactions* **61** p 754–64

Effect of strain rate on strength of unidirectional CFRP

A M Dumansky[1,2], M A Alimov[1] and A V Terekhin[3]

[1] Mechanical Engineering Research Institute of the Russian Academy of Sciences, 4, Maly Kharitonyevskiy Pereulok, Moscow, 101990, Russia
[2] Bauman Moscow State Technical University, 5/1, 2nd Baumanskaya Street, Moscow, 105005, Russia
[3] State Research Center "ORPE Technologiya" named after A.G. Romashin, 15, Kievskoe Shosse, Obninsk, Kaluga Region, 249031, Russia

aldumans@mail.ru

Abstract. An analysis of the anisotropy of failure stress envelopes for unidirectional carbon fiber reinforced plastic at different strain rates, including dynamic loading, is carried out. Based on the study of transverse compressive and share strengths, a peak of the failure stress envelopes is observed, which is described by using the parabolic function approximation. It is shown that with an increase in the strain rate, not only an increase in strength but also a shift of the peaks to the area of higher transverse compressive stresses are observed.

1. Introduction

In recent times, more and more publications related to the study of sensitivity of polymeric CFRPs to strain rate and other time-varying loads have appeared [1-5]. To a large extent, this is associated with the study of mechanisms of resistance to cracking, ability to absorb energy which does not cause degradation of material properties and, as a result, to the problem of ensuring the reliability and service life of critical structural components made of polymer composites. In particular, researches continue despite the fact that some airframe components made of polymeric composites have been launched into mass production.

A review of the regularities of deformation and fracture of polymer composite materials [6] notes that, in the overwhelming majority of cases, there is an increase in Young's modulus and strength with an increase in strain rate. Only in two cases relating to glass/epoxy and glass/polyester composites, the strength decreases with increasing strain rate. Most studies were carried out in a wide range of strain rates, including dynamic loading, which were realized using a split-Hopkinson pressure bar (SHPB). It should be noted that when studying elastic and rheological properties, a weak sensitivity of Poisson's ratio to strain rate is observed, which is explained by fiber properties.

A detailed study of the regularities of mechanical behavior of IM7/8552 unidirectional carbon fiber reinforced plastic was carried out [7, 8]. It is noted [7] that due to the effect of transverse compressive stress, modification of the known failure criteria of unidirectional composites, such as Tsai-Wu, Tsai-Hill, Hashin-Rotem, etc., is required. The sensitivity of mechanical properties to strain rate requires taking into account the influence of the viscoelastic properties of the matrix under transverse stresses and in-plane shear. It is also noted that an increase in resistance to cracking with an increase in strain rate is observed under dynamic loading. Using high speed camera images made it possible to analyze and measure the fracture angles. The phenomenon of fiber rotation due to deformation of specimens was observed under off-axis loading [7]. An analysis of the strain-rate curves of off-axis specimens shows the sensitivity to strain rate as well as the appearance of nonlinearity in such curves. Along with this, any change in Young's modulus in the direction of reinforcement due to strain rate can be neglected. Attempts [7, 9] were made to use the well-known Puck criterion [10] which generalizes the well-known Mohr-Coulomb law, allowing the effect of normal stresses on shear stresses in the fracture plane to be taken into account. A consequence of this approach is the ability to describe a shear strength peak in the presence of normal stresses. In this case, the well-known Hashin-Rotem criterion with Sun's correction takes the following form

$$\left(\frac{\sigma_2}{\sigma_2^-}\right)+\left(\frac{\tau_{12}}{\tau_{12}^*-\eta\sigma_2}\right)=1,$$

where η is an experimentally determined parameter that allows the effect of friction forces caused by normal stresses to be taken into account, σ_2^- is the transverse compressive strength, and τ_{12}^* is the in-plane shear strength.

The lamina plastic flow criterion [11] representing a simplified one-parameter version of the anisotropic theory of plasticity of a lamina caused by a certain combination of stresses reaching some yield point is widely used. Adding two more additional parameters to such theory made it possible to expand the field of application of the theory to the case of taking into account the effect of strain rate as well as describing the viscoplastic deformation of unidirectional CFRPs [12-14]. A study of the effect of fiber orientation on the mechanical properties of unidirectional IM7/8551-7 CFRP specimens under loading at various strain rates [15] shows that the dependence of the ultimate strain on the direction of fiber orientation is non-monotonic. In the range from 45° to 75°, peaks of the ultimate strain values were observed, and the relationship between the strength values under quasi-static and dynamic loading was determined using a correction factor being an off-axis angle polynomial and including the natural logarithm of strain rate as a multiplier. A similar method was used in [16]. The effect of fiber rotation in IM7/8552 angle-ply laminates under tensile and compressive loading in the area of large strains was studied in [17]. Model constructions of constitutive relations using the linear viscoelasticity relations based on the Maxwell bodies connected in parallel with an elastic spring was performed in [18] where the stiffness matrix components are considered as functions of time in a hereditary form. A criterion taking into account the predominance of one or another fracture mechanism and making it possible to describe the intra-laminar strength peak in the area of transverse compression was proposed in [19].

The results of studying the calculated and experimental characteristics of deformation and failure of unidirectional and layered composite materials under quasi-static loading are summarized in [20] where the predictions using different models and mechanical properties of unidirectional specimens [21] were compared with test results. A criterion for the adequacy of the models was the proximity of predicted and experimental data. A data analysis is given in [22] where a significant scatter of the predicted data is noted and the number of input parameters introduced into the proposed theories is discussed. Taking into account the influence of variable loads is, to some extent, a continuation of the above-mentioned works providing for the variation of parameters that allow the effect of time and physical nonlinearity of properties to be described. In particular, during the rate deformation of specimens made of unidirectional material, the dependence of the stress-strain curves on rate and the physical nonlinearity are observed. An analysis of the regularities under quasi-static loading and the effect of the strain rate on the rheological properties are considered in [23, 24].

2. Strain rate dependence of failure envelopes

Let us consider the application of the well-known Tsai-Wu failure criterion in relation to the fracture of unidirectional material loaded at an off-axis angle (see figure 1)

$$\left(\frac{1}{\sigma_2^+}-\frac{1}{\sigma_2^-}\right)\sigma_2+\frac{\sigma_2^2}{\sigma_2^+\sigma_2^-}+\frac{\tau_{12}^2}{\tau_{12}^{*2}}=1,$$

where $\sigma_2^{\pm}$ are the transverse tensile and compressive strengths, τ_{12}^* is the in-plane shear strength. The Tsai-Wu criterion describes a failure envelope in the form of an ellipse but does not allow of describing the experimentally observed strength peak associated with an increase in shear strength in the presence of transverse compressive stresses. Similarly, other criteria, with the exception of the Sun criterion, do not allow of taking this factor into account. The Northwestern University (NU) criterion [1, 2] allows the strength peak to be described by associating it with a change in the nature of failure, namely, with an increase in compressive stresses, when the shear nature of failure is replaced by the prevalence of compressive stresses.

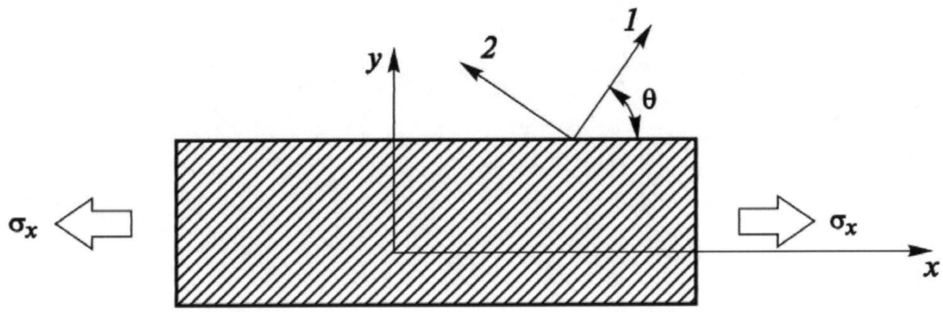

Figure 1. Schematic illustration of off-axis loading: directions 1 and 2 are the major orthotropy axes, θ is off-axis angle.

Let us introduce a parabolic dependence of the shear strength on transverse stresses

$$\tau_{12}(\sigma_2)=a\sigma_2^2+b\sigma_2+c$$

Let us propose an approach based on a quadratic least squares approximation of the failure stress envelopes for IM7-8552 unidirectional CFRP specimens tested at different off-axis angles and at three strain rates: quasi-static, intermediate, and dynamic [2]. The test results digitized from the figures in [2] were approximated by the following curves

$$\tau_{12}=-0.0039\sigma_2^2-0.9046\sigma_2+88.98$$

$$\tau_{12}=-0.0031\sigma_2^2-0.946\sigma_2+104.64$$

$$\tau_{12}=-0.0029\sigma_2^2-0.961\sigma_2+121.68$$

The first equation describes the failure envelope corresponding to quasi-static loading at the strain rate equal to 10^{-4} s^{-1}. The second equation corresponds to the intermediate strain rate equal to 1 s^{-1}, and the third to dynamic loading at the strain rate $\dot{\varepsilon}$ = 800 s^{-1}. R-squared values (R^2) for the respective strain rates were 0.9144, 0.8529 and 0.8806 accordingly.

An analysis of the experimental data shows that the failure envelopes do not intersect each other and the curves corresponding to higher strain rates lie higher (see figure 2). In addition, it can be seen that the parabola peaks shift to the left with increasing strain rate.

The resulting curves allow the peaks of the failure envelopes under shear and transverse compressive stress conditions to be described. It is obvious that, when $\sigma_2=0$, the curves make it possible to estimate the values of in-plane shear strength, i.e. in the absence of transverse stresses. For the corresponding strain rate, the shear strength is equal to the free term in the respective parabola equation. For example, the in-plane shear strength equal to 90 MPa is given in [21], while the free term in the respective failure envelope equation is 88.98 MPa. The maximum shear strength values under compressive stress conditions are $\tau_{\max} = -\dfrac{b^2}{4a} + c$ with $\sigma_2 = -\dfrac{b}{2a}$ as a parabola axis of symmetry. It should be noted that there are some regularities in changing the parabola parameters depending on the strain rate. With an increase in the strain rate, these changes, on the one hand, characterize the shift of the parabola peak to the left and, on the other hand, raise the parabola up.

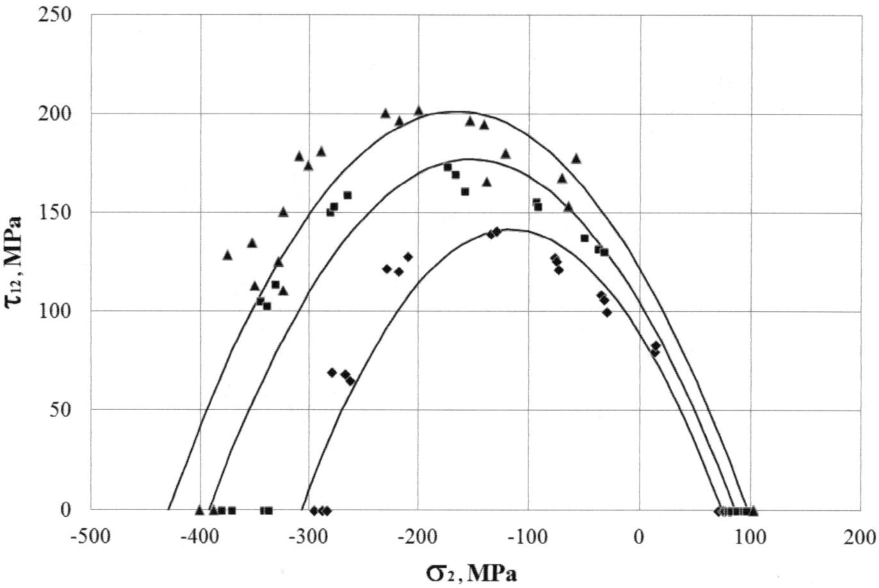

Figure 2. Strain-rate dependent failure envelopes for IM7-8552 unidirectional composite: ▲ - 800 s^{-1}, ■ - 1 s^{-1} , ♦ - 10^{-4} s^{-1}.

3. Conclusions

To describe the strength of unidirectional CFRP, a simple parabolic dependence is proposed, which make it possible to describe the failure envelopes at different strain rates as well as the experimentally observed peak in the shear strength. The free terms in the parabola equation allow us to estimate the in-plane shear strength.

References

[1] Daniel I 2014 Failure of composite materials under multi-axial static and dynamic loading *Procedia Engineering* **88** pp 10-17

[2] Schaefer J D, Werner B T and Daniel I M 2018 Strain-rate-dependent criteria for composite laminates: Applications of the Northwestern failure theory to multiple material systems *Mechanics of Composite and Multi-functional Materials* (*Conf. Proc. of the Society for Experimental Mechanics Series* vol 6) ed P R Thakre et al chapter 19 pp 187-196 DOI 10.1007/978-3-319-63408-1_19

[3] Koerber H, Xavier J and Camanho P P 2010 High strain rate characterization of unidirectional carbon-epoxy IM7-8552 in transverse compression and in-plane shear using digital image correlation *Mechanics of Materials* **42** pp 1004-19

[4] Sun T C and Tao J 1998 Prediction of failure envelopes and stress/strain behavior of composite laminates *Composite Science and Technology* **58** pp 1125-36

[5] Ploekl M, Wolfhart M, Kuhn P and Koerber H 2017 A dynamic test methodology for analyzing the strain-rate effect on the longitudinal compressive behavior of fiber-reinforced composites

Composite Structures **180** pp 429-38

[6] Jacob G C, Starbuck J M, Fellers J F, Simunovic S and Boeman R G 2004 Strain rate effects on the mechanical properties of polymer composite materials *J. of Applied Polymer Science* **94** pp 296-301

[7] Koerber H, Havier J and Camanho P P 2010 High strain rate characterization of unidirectional carbon-epoxy IM7-8552 in transverse compression and in-plane shear using digital image correlation *Mechanics of Materials* **42** pp 1004-19

[8] Kuhn P, Ploeckl M and Koerber H 2015 Experimental investigation of the failure envelope of unidirectional carbon-epoxy composite under high strain rate transverse and off-axis loading *EPJ Web of Conf.* **94** DOI: 10.1051/epjconf/20159401040

[9] Davila C G, Camanho P P and Rose C A 2005 Failure criteria for FRP laminates *J. of Composite Materials* **39** No 4 pp 323-43

[10] Puck A and Schűrmann H 1998 Failure analysis of FRP laminates by means of physically based phenomenological models *Composite Science and Technology* **62** pp 1633-62

[11] Sun C T and Chen J L 1998 A simple flow rule for characterizing nonlinear behavior of fiber composites *J. of Composite Materials* **23** pp 1009-20

[12] Bing Q and Sun C T 2005 Modeling and testing strain rate-dependent compressive strength of carbon/epoxy composites *Composite Science and Technology* **65** pp 2481-91

[13] Cho J, Fenner J, Werner B and Daniel I M 2010 A constitutive model for fiber-reinforced polymer composites *J. of Composite Materials* DOI: 10.1177/0021998310371547

[14] Thiruppukuzhi S V and Sun C T 1998 Testing and modeling high strain rate behavior of polymeric composites *Composites Part B* **29B** pp 536-46

[15] Vinson J R and Woldesenbet E 2001 Fiber orientation effects on high strain rate properties of graphite/epoxy composites *J. of Composite Materials* **35** No 6 pp 509-21

[16] Raimondo L, Lanucci L, Robinson P and Curtis P T 2012 Modelling of strain rate effects on matrix dominated elastic and failure properties of unidirectional fibre-reinforced polymer-matrix composites *Composite Science and Technology* **72** pp 819-27

[17] Cui H, Thomson D, Pellegrino A, Wiegand J and Petrinic N 2016 Effect of strain rate and fibre rotation on the plane shear response of ±45 laminates in tension and compression tests *Composite Science and Technology* **135** pp 106-15

[18] Karim M R and Fatt M S H 2006 Rate-dependent constitutive equations for carbon fiber-reinforced epoxy *Polymer Composites* **27** pp 513-28

[19] Daniel I M, Luo J J, Shuber P M and Werner B T 2009 Interfiber/interlaminar failure of composites under multi-axial states of stress *Composite Science and Technology* **69** pp 764-71

[20] *Failure Criteria in Fibre Reinforced Polymer Composites: The World-Wide Failure Exercise* 2004 ed V J Hinton *et al* (Amsterdam: Elsevier) p 1269

[21] Kaddour A S, Hinton M J, Li S and Smith P 2007 *Instructions of the Third World-Wide Failure Exercise (WWFE-III)* Part A p 48

[22] Christensen R M 2019 The World Wide Failure Exercise II: Examination of results https://www.failurecriteria.com/theworldwidefail.html

[23] Dumansky A, Liu Hao and Alimov M 2019 Strain-rate effect on mechanical behavior of unidirectional carbon fiber reinforced plastic *Materials Today: Proc.* **19** pp 2377-81

[24] Dumansky A M, Liu Hao and Alimov M A 2019 Analysis of nonlinear behavior of uniderectional CFRP under quasi-static loading *Proc. 11th All-Russian Conf. on Test and Research of Properties of Materials "TestMat" (Physical and Mechanical Tests, Strength, Reliability, High Temperature Tests)* ed Slavin A V *et al* (Moscow: VIAM) pp 143-52

IOP Conf. Series: Materials Science and Engineering 1023 (2021) 012010 doi:10.1088/1757-899X/1023/1/012010

Multiscale modeling of damage and fracture of a composite overwrapped pressure vessel

N V Eremin

Krasnoyarsk Branch of the Federal Research Center for Information and Computational Technologies, Krasnoyarsk, Russian Federation

kaizoku813@gmail.com

Abstract. An approach to multiscale modeling of damage and fracture of a composite overwrapped pressure vessel based on experimental studies is presented. The proposed approach takes into account the interaction of mechanical properties and structural parameters of the composite material at each scale level and their joint interaction. A micromodel of the fracture of a composite material has been developed taking into account the structural features. Calculated mechanical and strength properties of unidirectional composite material are determined. A mesomodel of fracture of a layered composite material is presented for numerical three-point flexural tests. Experimental verification of the numerical mesomodel of straining and fracture of a layered composite material has been carried out. A comparative analysis of the experimental data and numerical modeling showed the quantitative and qualitative agreement of the results. A macromodel of damage and fracture of a composite overwrapped pressure vessel is presented. The diagrams of progressive damage in a composite shell depending on the mode of fracture are presented. An assessment of the strength of the vessel structure is given taking into account the initiation and evolution of damage in the composite shell.

1. Introduction

The design of a composite overwrapped pressure vessel (COPV) consists of a VT1-0 sealing titanium liner and a 9-layer composite shell based on IMS60 carbon fibers and an ED-I epoxy matrix. Modeling the fracture of structures using composite materials (CM) made by filament winding is a complex task.

Straining and fracture of a layered composite shell is a multi-stage process that covers various scale levels [1-6], where each level is interrelated and affects the general characteristics of mechanical properties. Three scale levels are considered: microlevel, mesolevel and macrolevel. When CM is subjected to mechanical loads, various fracture processes occur within all three scales simultaneously.

Analysis of literature sources led to the conclusion that in the development multiscale numerical models, the relationship with experimental studies is not always traced. In numerical simulation of fracture of a COPV, simplified models are used, which do not take into account the elastic plastic properties of the liner and contact. In addition, there are not many authors who take into account the design and technological features of CM and their influence on the stress-strain state of the COPV.

In this regard, for a better understanding of the behavior the straining and fracture of the COPV, it becomes necessary to development numerical models that take into account the processes of fracture at each scale level. The aim of this work was to development multiscale model of the fracture of a

COPV taking into account the design and technological features of CM. To achieve this goal, several tasks were solved:

- development of a micromodel of fracture of CM to determine the mechanical properties and ultimate strengths depending on the type of loading;
- experimental verification of the calculated mechanical and strength properties on the numerical mesomodel of fracture of a layered CM;
- development of a macromodel of damage and fracture of a COPV for studying the processes of straining and fracture depending on the applied pressure.

2. Micromodel of composite material fracture

In the CM mechanics, systems of joint interaction of structural elements are considered, taking into account micromechanical effects on the "fibers-matrix" scale [7]. Such models make it possible to predict effective mechanical properties, inhomogeneous stress and strain fields, and also to simulate the fracture of CM as a multistage process.

Using the ANSYS Mechanical APDL finite element analysis, a 3D numerical micromodel of CM was developed, which includes a number of features: generation of random arrangement of carbon fibers inside the matrix; modeling of random distribution of mechanical properties; modeling fracture of structural elements.

Based on the studies in [8], the fibers in the CM are located randomly and even with large distances between them, which makes it necessary to model a variant with a random arrangement of fibers inside matrix. In addition, most carbon fiber-reinforced CM have a high variation of mechanical properties [9]. The coefficient of variation of the Young's modulus is 0.22, and for the ultimate tensile strength is 0.25. The model of the scatter of mechanical properties is based on the division of each fiber into sectors with different values of mechanical properties, which are described according to the normal distribution.

The fracture of structural elements of the CM is modeled by reducing the stiffness matrix of the elements by the degradation coefficient $D=1\cdot10^{-6}$. At each stage of loading, an analysis of stresses in the elements is carried out, if the element does not fulfill the strength condition, then it fracture. In these studies, the criterion of maximum stress was used depending on the type of loading.

A numerical micromodel of a unidirectional CM with a chaotic arrangement of fibers, in which the fiber and matrix are isotropic materials is are presented in figure 1. Geometrical parameters of the model: carbon fiber percentage $V_f = 70\%$, fiber diameter $d_f = 5$ μm, fiber length $L = 650$ μm, side of a square B =30 μm.

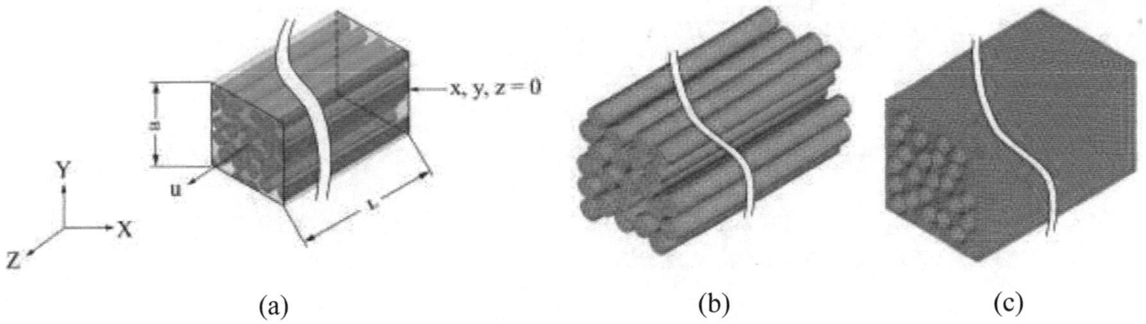

(a) (b) (c)

Figure 1. Numerical micromodel of fracture of a unidirectional CM: (a) tensile design scheme; (b, c) finite element of model of structural elements of CM.

Mechanical and strength properties of structural elements in CM are presented in table 1.

Table 1. Mechanical and strength properties of structural elements CM.

Characteristic	Carbon fiber IMS60	Epoxy resin ED-I
Young's modulus, MPa	290000	3000
Tensile strength, MPa	5800	50
Poisson's ratio	0.28	0.32

On the basis of the developed micromodel, a multivariate numerical experiments of CM were carried out and the main of mechanical properties and ultimate stress values were determined depending on the type of loading: under tension (X_t) and compression (X_c) along the direction of reinforcement; under tension (Y_t) and compression (Y_c) across the direction of reinforcement; shear (S_l). The fracture process of CM under the action of a tensile load in the direction of the reinforcement is presented in figure 2.

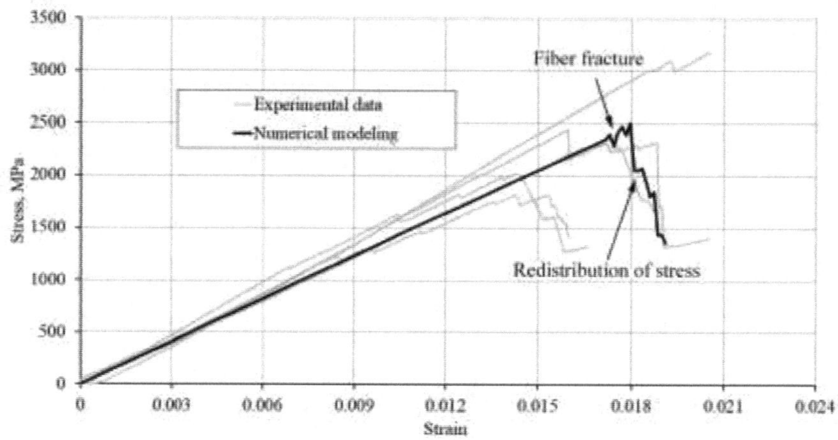

(a)

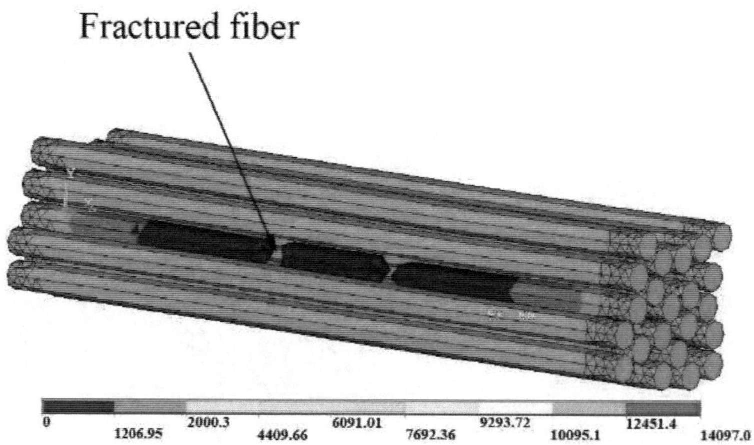

(b)

Figure 2. Numerical micromodel of fracture of a unidirectional CM: (a) strain curve of CM under tension; (b) fiber fracture simulation process.

The calculated mechanical and strength properties CM are presented in table 2 and 3.

Table 2. Mechanical properties CM.

Characteristic	E_1, MPa	$E_2 = E_3$, MPa	$\nu_{12} = \nu_{13}$	ν_{23}	$G_{12} = G_{13}$, MPa	G_{23}, MPa
Value	129000	5800	0.32	0.45	3400	3800

Table 3. Ultimate strength properties CM.

Characteristic	X_t, MPa	X_c, MPa	Y_t, MPa	Y_c, MPa	S_l, MPa
Value	2334	-969	50	-100	50

Thus, the calculated values of mechanical properties and ultimate strengths were determined with a known set of data on each structural element of the CM.

3. Mesomodel of fracture of a laminated composite material

The mesomodel consists of a package of differently oriented composite tapes. Continuum damage model (CDM) was used to simulate fracture at the meso and macro levels. The model is based on a continuous process of degradation of the mechanical properties of the material due to the accumulation of damage [10]. Material degradation spreads throughout the model according to the failure criteria as long as the structure is able to redistribute the load. This causes to the use of CDM at the meso and macro scales, since at the microlevel, fracture is modeled by a sharp decrease in the stiffness matrix of the element and its exclusion from the calculation, which in turn not allow correctly displaying the process of straining and fracture of the CM.

To predict progressive damage using CDM, it is necessary to determine 8 parameters: 4 values of energy dissipated per unit area depending on the fracture mode; 4 viscous damping coefficients depending on the fracture mode.

Energy dissipated per unit area is determined by the equation:

$$G_c = \int_0^{u_{eq}^f} \sigma_{eq} du_{eq} = \frac{1}{2}\sigma_{eq}^f u_{eq}^f, \tag{1}$$

where σ_{eq}, u_{eq} – equivalent stress and displacement, respectively; σ_{eq}^f, u_{eq}^f – ultimate equivalent stress and displacement, respectively.

Using equation (1) and the results obtained at the microlevel, the energy presented in table 4 was calculated.

Table 4. Energies parameters of CM.

Characteristic	G_c^{Xt}, N·mm^{-1}	G_c^{Xc}, N·mm^{-1}	G_c^{Yt}, N·mm^{-1}	G_c^{Yc}, N·mm^{-1}
Value	23.34	5.82	0.325	0.7

Viscous damping coefficients for the various modes were taken from the [11] and presented in table 5.

Table 5. Energies parameters of CM.

Characteristic	ζ_t^f	ζ_c^f	ζ_t^m	ζ_c^m
Value	$1 \cdot 10^{-3}$	$1 \cdot 10^{-3}$	$5 \cdot 10^{-3}$	$1 \cdot 10^{-3}$

The initiation of damage is determined using failure criterion. In these studies, the Hashin failure criterion [12] was used to determine the mods of fracture. Damage parameter *d* for each type of fracture is calculated by equivalent displacements. Damage variables increase gradually based on the energy amounts dissipated for the various damage modes:

$$d = 1 - \frac{u_{eq}^{0}(u_{eq}^{f} - u_{eq})}{u_{eq}(u_{eq}^{f} - u_{eq}^{0})}, \tag{2}$$

where u_{eq}^{0} – equivalent displacements corresponding to initiation of fracture.

Due to the initiation of damage in the material, the stiffness in the elements decreases. The damaged elastic matrix is defined as follows:

$$[C]_d = \frac{1}{A}\begin{bmatrix} (1-d_f)E_f & (1-d_f)(1-d_m)\nu_{21}E_f & 0 \\ (1-d_f)(1-d_m)\nu_{12}E_f & (1-d_m)E_m & 0 \\ 0 & 0 & A(1-d_s)G \end{bmatrix}, \tag{3}$$

where $A = 1 - \nu_{12}\nu_{21}(1-d_f)(1-d_m)$; E_f, E_m, G – undamaged elastic and shear modulus; d_f, d_m, d_s – damage parameters depending on the mode of fracture.

Experimental verification of the numerical mesomodel of straining and fracture of a layered CM has been carried out. Plane specimens of a 9-layer CM were modeled at various reinforcement angles relative to the central axis (0, 45 and 90 degrees) for numerical three-point flexural tests (figure 3). The model used mechanical properties and ultimate strength values calculated on the micromodel.

Average geometric dimensions of specimens and models: number of layers $n = 9$; layered thickness $h_{ply} = 0.34$ mm; thickness $h = 3.06$ mm; width $b = 20$ mm. Radius of curvature for different reinforcement angles: $R_0 = 506$ mm; $R_{45} = 504$ mm; $R_{90} = 236$ mm.

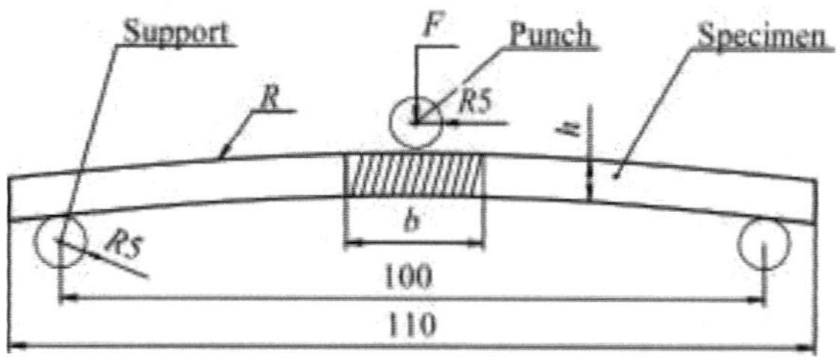

(a)

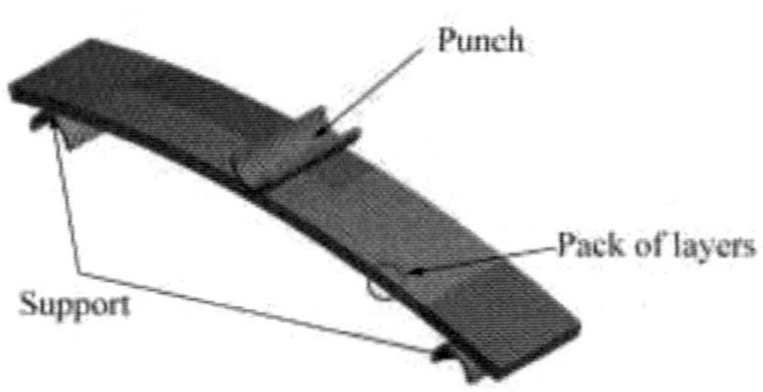

(b)

Figure 3. Design scheme (a) and finite element model (b) for three-pint flexural test.

Based on the results of the numerical modeling of the CM for three-point flexural tests, “load-deflection” diagrams were constructed for each specimen (figure 4).

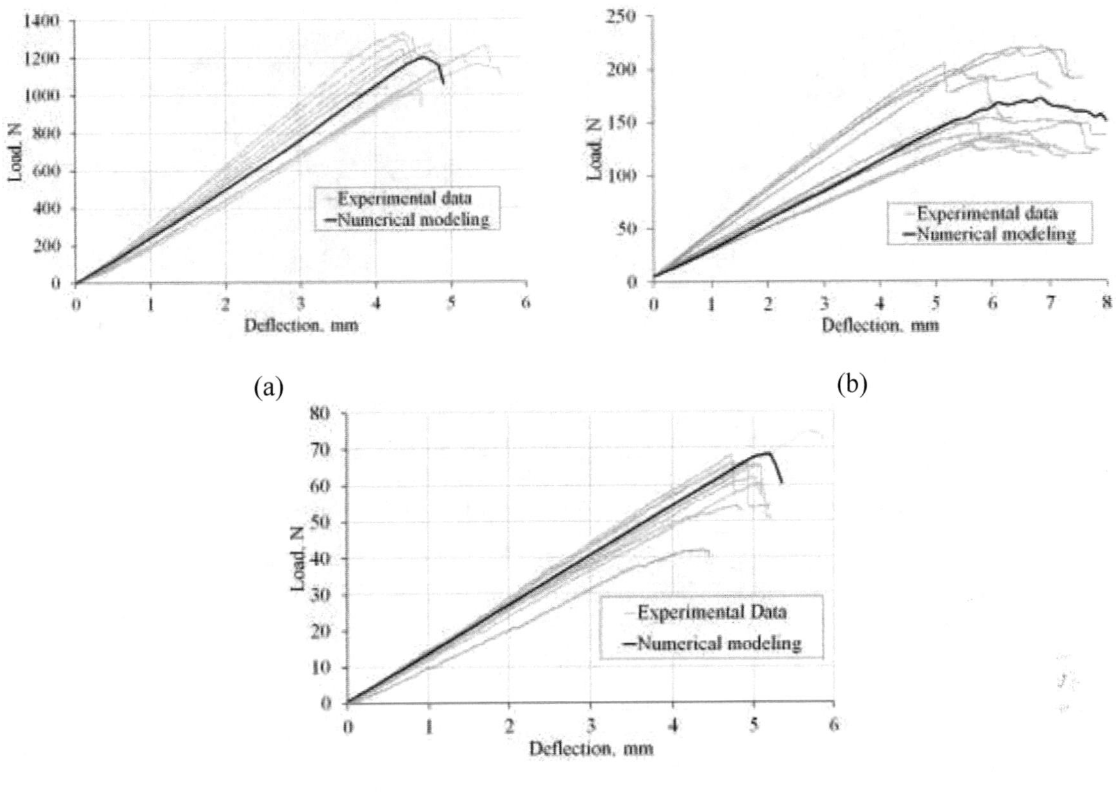

Figure 4. "Load-deflection" diagrams of laminated CM at different reinforcement angles: (a) 0°; (b) 45°; (c) 90°.

Comparative analysis of experimental data [8] and numerical modeling showed the qualitative and quantitative correspondence of the straining and fracture process of specimens with different reinforcement angles. The considered numerical model, which is determined by the mechanical and strength properties of the CM, as well as damage and fracture modeling using CDM, is applicable for COPV modeling and allows to reliably estimate its strength.

4. Macromodel of the fracture of a composite overwrapped pressure vessel

In most cases, the analysis of the stress-strain state of the COPV is performed using axisymmetric finite elements or three-dimensional elements of a layered shell [13, 14]. Initially, the cross-section of the liner and the composite shell was determined on the basis of theoretical relations for calculating the thickness [15], and then a volumetric model of the repeating sector was built by rotating around the axis. Due to the technological features of winding the composite tape on the liner, the thickness of the composite shell is not constant, it continuously increases from the equator to the pole hole. Composite shell formed by two-zone filament winding.

COPV geometrical dimensions: shell radius $R = 506$ mm; shell height from equator $Z = 281$ mm; pole hole radius $r_1 = 35$ mm and zone of "hump" radius $r_2 = 95$ mm. The design scheme and finite element model are presented in figure 5.

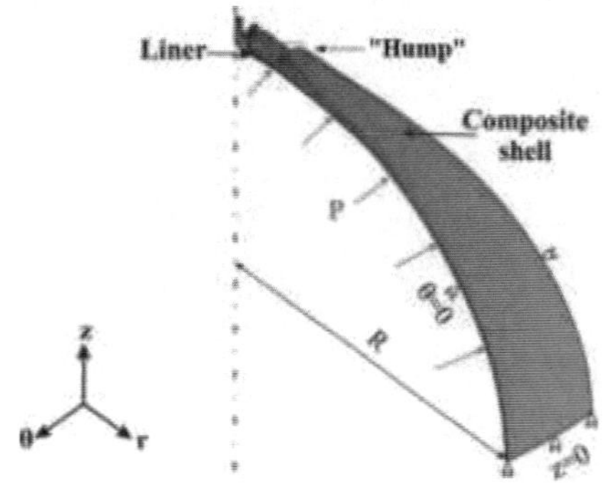

(a)

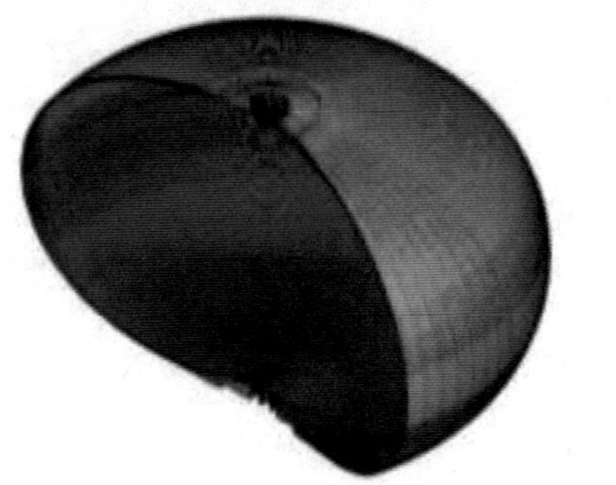

(b)

Figure 5. Numerical macromodel of COPV: (a) design scheme; (b) finite element model.

The mechanical properties of elastic-plastic titanium VT1-0 are presented in table 6.

Table 6. Mechanical properties of titanium liner.

Characteristics	*E*, MPa	μ	σ_{yield}, MPa	σ_{US}, MPa
Value	110000	0.32	380	430

The pressure was applied to the inner surface of the liner with a constant increase in value. The contact between the liner and the composite shell was set with a friction coefficient of 0.2.

Based on the performed numerical calculations, the curves of equivalent strains of the structural elements of the COPV were constructed, taking into account the initiation and evolution of damage in the composite shell (figure 6). Based on the figure, it can be seen that the structure loses its bearing capacity at a pressure of 20.25 MPa.

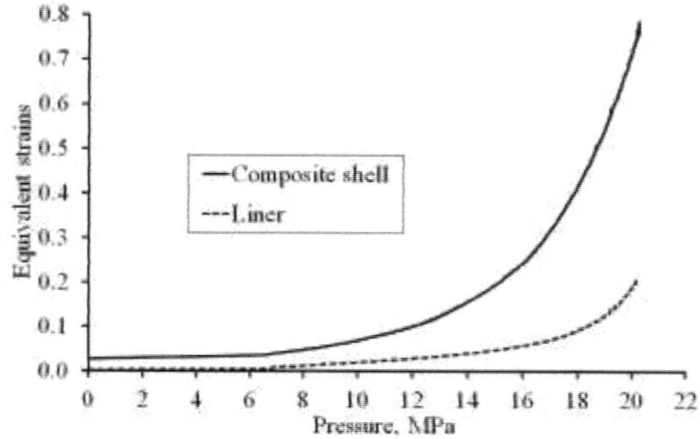

(a)

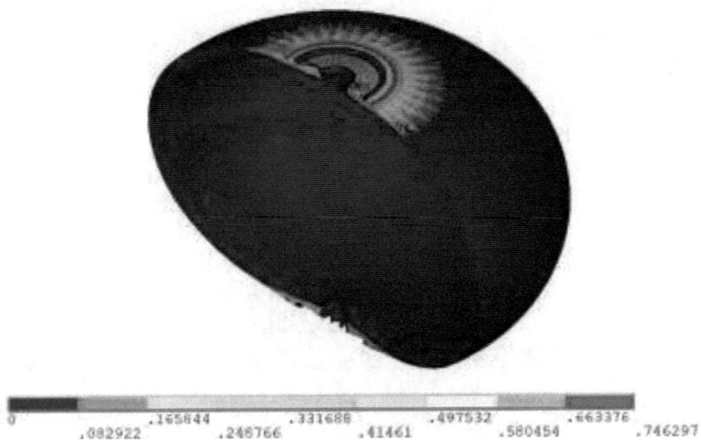

(b)

Figure 6. Curves of equivalent strains of structural elements of COPV (a) and the distribution of total damage in the composite shell at a pressure of 17.5 MPa (b).

The distribution of damage in the composite shell under different modes of fracture is presented in figure 7. The distribution of damage at $P / P_{max} = 0.4$ corresponds to the initiation of damage in the composite shell, and $P / P_{max} = 1$ to its end. Damage level is rated in the range from 0 (intact element) to 1 (fractured element).

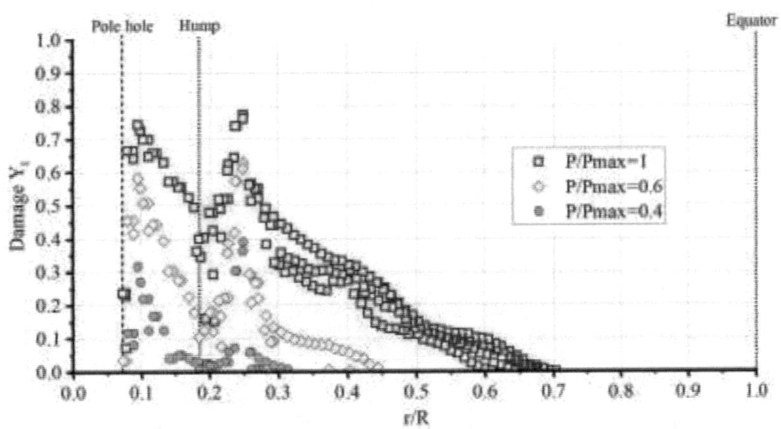

(a)

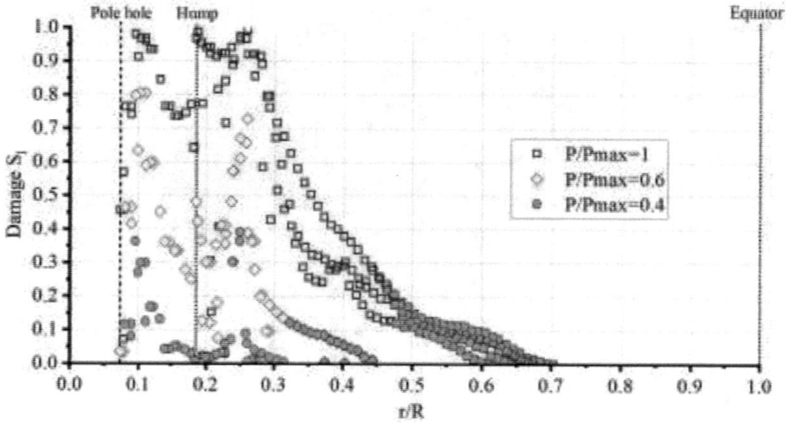

(b)

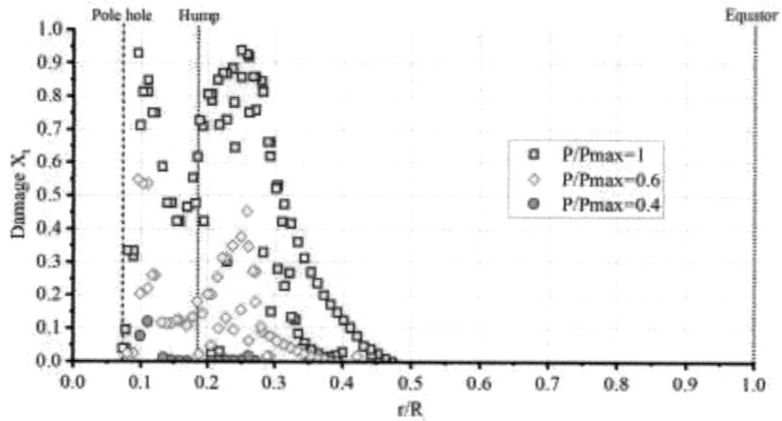

(c)

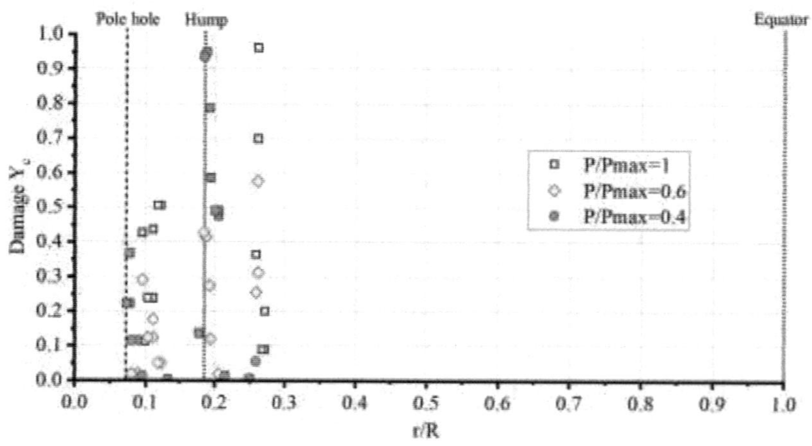

(d)

Figure 7. Distribution of damage depending on the position of the radius r of the composite shell: (a) matrix tension; (b) matrix shear; (c) fiber tension; (d) matrix compression.

Due to the lower strength of the CM in the transverse direction, damage to the matrix initially occurs in the area of pole hole and the "hump" zone (figure 7 (a)). As the load increases, damage grows along the direction of the equator. In addition, shear damage to the matrix occurs in the layers (figure 7 (b)), resulting in tensile damage in the carbon fibers (figure 7 (c)). The initiation of damages in the fibers during compression is not observed, and in the matrix, they are insignificant (figure 7 (d)). The circumstance is explained by the fact that behavior of the composite tape upon application of internal pressure is close to stress state of a unidirectional CM under tension along the direction of reinforcement.

When high damage occurs in the CM, extensive plastic strains occur in the liner in the area of the flange. In this area, a high concentration of local strains is observed, caused by a change in the stiffness of the material due to an increase in the thickness of the flange in comparison with the bottom membrane. Parallel to this, the stiffness in the layers is significantly reduced, which leads to a loss of the COPV load-bearing capacity.

Under the action of internal pressure in the COPV, several stages of deformation and destruction can be distinguished: elastic straining of the CM; elastic-plastic straining of the liner; initiation and evolution of damage in the CM; liner critical straining; local fracture of the composite shell and liner; loss of the COPV load-bearing capacity.

5. Conclusions

An approach to multiscale modeling of a composite overwrapped pressure vessel based on experimental studies is presented. Developed multilevel models of fracture of CM taking into account design and technological features.

A micromodel of CM fracture is presented, taking into account random structural features. The proposed model allows one to determine the effective mechanical and strength properties with a known set of data on the structural elements of CM.

An experimental verification of the obtained mechanical and strength properties, as well as parameters characterizing damage and fracture at the meso and macro levels, has been carried out. Comparative analysis showed qualitative and quantitative agreement of the results.

The COPV micromodel has been developed, which takes into account the technological and design features of the CM. Using the CDM, the developed numerical model allows for reliable tracking of straining processes, as well as damage and fracture in the composite shell. The determined limiting pressure for the considered configuration of the COPV, which is 20.25 MPa.

Thus, the developed combined approach makes it possible to track the processes of straining and fracture of CM at each scale level, and to assess the strength of the COPV at design stage.

References

[1] Talreja R 2016 Multiscale modeling of failure in composite materials *Proc. of the Indian National Science Academy* **82**(2) pp 173-81

[2] Littlefield A, Macri M, Smith L and Root J 2018 Multiscale modeling a composite pressure vessel subjected internal pressure *Conf.: CMAX 2018* p 17

[3] Vignoli L L and Savi M A 2017 Multiscale failure analysis of cylindrical composite pressure vessel: A parametric study *Latin American J. of Solids and Structures* **15**(11) p 20

[4] Min-Gu Han and Seung-Hwan Chang 2015 Failure analysis of a type III hydrogen pressure vessel under impact loading induced by free fall *Composite Structures* **127** pp 288-97

[5] Liu P F, Chu J K, Hou S J and Zheng J Y 2012 Micromechanical damage modeling and multiscale progressive failure analysis of composite pressure vessel *Composite Material Science* **60** pp 137-48

[6] Mayes J S and Hansen A C 2001 Multicontinuum failure analysis of composite structural laminates *Mechanics of Composite Materials and Structures* **8**(4) pp 249-62

[7] Haitao Liu, Jie Lin, Yazhou Sun and Jinyang Zhang 2019 Micro model of carbon fiber/cyanate ester composites and analysis of machining damage mechanism *Chinese J. of Mechanical Engineering* **32**(1) pp 52-63
[8] Eremin N V 2018 Analysis of microstructure of layer polymer composite material of metal composite overwrapped pressure vessel *Siberian J. of Science and Technology* **19**(2) pp 346-54
[9] Moskvichev E V and Eremin N V 2017 Evaluation of mechanical properties and thickness of composite shell of metal-composite high-pressure vessel *Deformation and Fracture of Materials* **12** pp 40-44
[10] Liang Wang, Chuanxiang Zheng, Hongying Luo, Shuang Wei and Zongxin Wei 2015 Continuum damage modeling and progressive failure analysis of carbon fiber/epoxy composite pressure vessel *Composite Structures* **134** pp 475-82
[11] Barbero E J and Shahbazi Mehdi 2017 Determination of material properties for ANSYS progressive damage analysis of laminated composites *Composite Structures* **176** pp 768-79
[12] Hashin Z and Rosen B W 1964 Composite cylinder assemblage (CCA) *J. of Applied Mechanics* **31** p 223
[13] Pramod R, Krishnadasan C K and Shanmugam N Siva 2019 Design and finite element analysis of metal-elastomer lined composite overwrapped spherical pressure vessel *Composite Structures* **224** p 18
[14] Wu Q G, Chen X D, Fan Z C and Nie D F 2015 Stress and damage analyses of composite overwrapped pressure vessel *Proc. Engineering* **130** pp 32-40
[15] Eremin N V and Moskvichev E V 2017 Experimental verification of the relationships for calculating the composite shell thickness of metal composite overwrapped pressure vessel *Composite Materials Constructions* **147**(3) pp 3-7

An optimal experiment planning in the problem of longevity evaluation at crack propagation stage

I V Gadolina[1], A N Romanov[1], A A Bautin[2] and L Papic[3]

[1]Mechanical Engineering Research Institute of the Russian Academy of Sciences (IMASH RAN), Moscow, Russia
[2] TsAGI, Zhukovsky, Russia
[3] DQM Research Center, Chachak, Serbia

gadolina@mail.ru

Abstract. A large number of factors influencing the fatigue crack rate are traditionally studied separately. In this case, their interactions are not taken into account, and the experiment does not possess the property of optimality. The article describes the authors' vision of the problem, how it would be possible to approach the solution of the research problem from the point of view of the theory of optimal experiment planning. Initially, it is proposed to consider three basic factors, namely: the influence of loading asymmetry, loading frequency and sequence of overloads. A full factorial experiment with varying factors at extreme levels is considered. The formalization of the factors describing the sequence of overloads is proposed.

The second stage of fatigue - namely, the stage of fatigue crack propagation, traditionally attracts a lot of attention of scientists and engineers. It is primarily due to the increasingly widespread use of the concept of condition-based operation, i.e. when the object continues to operate despite the presence of a macrocrack. The task of assessing the survivability is also relevant when extending the guaranteed resource of the other mechanical engineering products.

Method

The problem of estimating the rate of crack growth is far from being trivial. There are many models for different materials and loading conditions [1-4]. The type of structural element is also important. An additional contribution to the variation of experimental results (as well as calculation results) is due to the random character of loading. It is known that overloads can slow down the rate of fatigue crack propagation. The scientists associate this phenomenon with the formation of a plastic zone, which leads to crack growth time lapse. Many calculation methods are based on the assessment of this influence. In [5], based on numerical simulations using the special method of random sequences imitation [6] and the estimate of the crack growth rate according to Willenborg [4], the effect of the sequence of overload peaks was investigated.

There is a wealth of experimental data on fatigue crack propagation. Perhaps the primary task should be to generalize them. On the other hand, it is interesting to outline further directions of research. For a systematic approach, it is advisable to formulate the problem of planning an optimal experiment to study the phenomenon of crack growth rate in metals. In this case, it is important to determine the choice of response- a function that optimally characterizes the stage of crack development.

Following the theory of optimal experiment planning [7], it is recommended to consider the factors possibly influencing the response. If three factors are examined, they are normalized, and the experiment space looks like a unit cube.

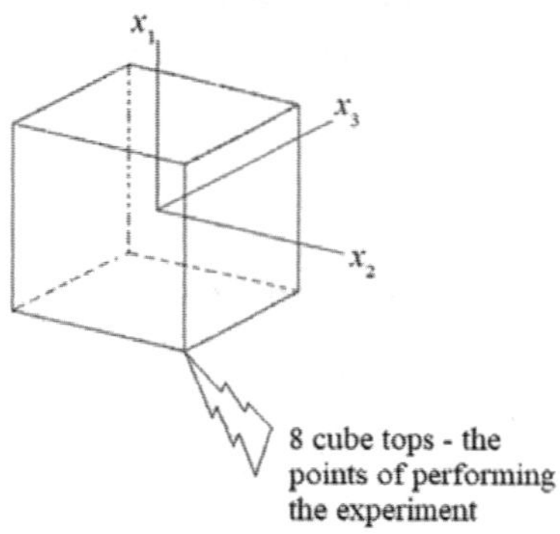

Figure 1. Single cube in factor space.

Having carried out a full factorial experiment (figure 1), one can get an idea of the main trends of influence and outline combinations of factors for further research. Long-term efforts of scientists have led to an understanding of WHICH factors can be chosen as determining factors in the task of crack propagation: the loading asymmetry (X_1), the loading frequency (X_2) and sequence of peaks (X_3). The experimental data for factors X_1 and X_2 and the assignment of the boundaries of their variation would not present serious difficulties. Concerning the factor X_3, there is no formalization so far. Moreover, it would be desirable to characterize the factor as a quantitative one, and not just "yes" - "no". The latter definition is only justified to encode the quality factors.

Characterization of the sequence of loading peaks

In [5], in a numerical experiment, the duration of the propagation of a fatigue crack in an aluminium plate was estimated for random processes simulated based on TWIST using the original method of modelling the sequence [6]. The sequence of peaks in the processes varied. Other parameters remained unchanged. Since the evaluation by the Willenborg [4] method takes into account the sequence of peaks (which is essential in reality), the calculated durability turned out to be different. In figure 2, one can see that the difference in durability appeared to be significant (2.5 times), which was ensured by the proper selection of loading processes. On the base of results, presented in figure 2, two marginal processes were chosen: the "Light" and the "Severe" ones, which were further analyzed.

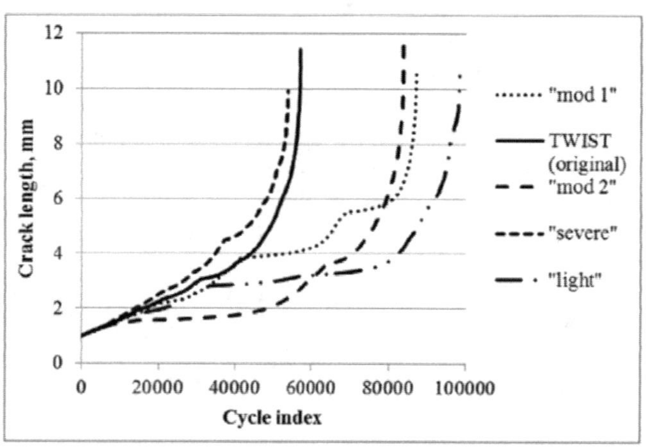

Figure 2. Calculated longevities (crack lengths) for varied processes.

Figure 3 shows the maxima of the light and severe processes.

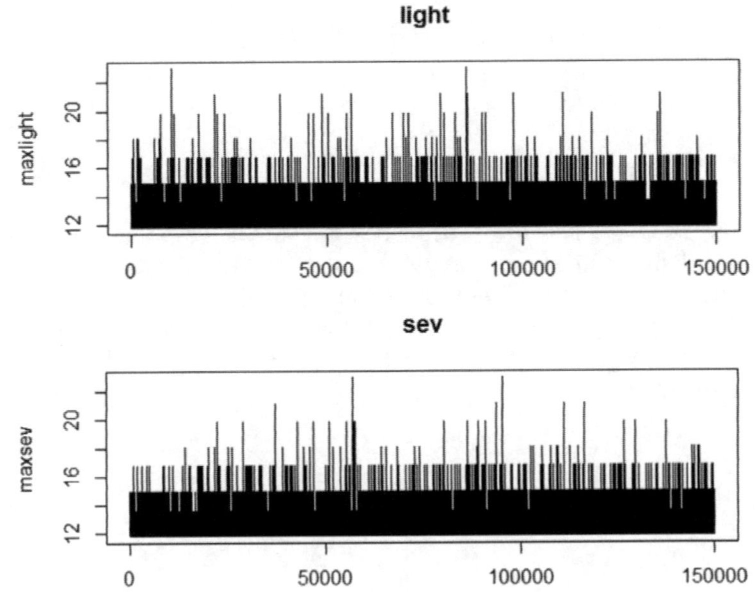

Figure 3. Maximums of two contrast processes.

The cumulative numbers of overloads over 17 kg·cm^{-2} for the shown in figure 3 "Light" and "Severe" processes are shown in figure 4.

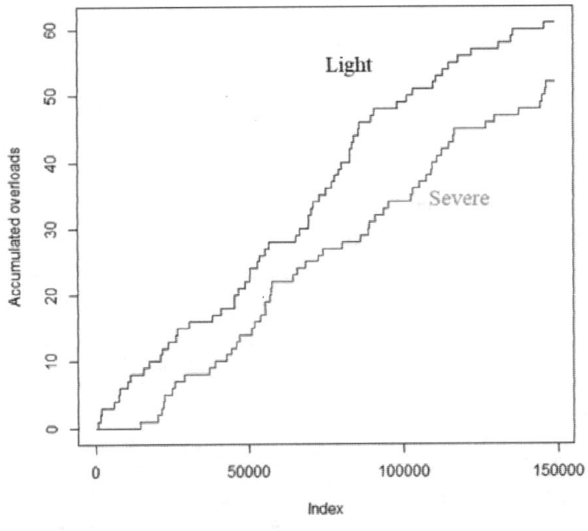

Figure 4. Cumulative numbers of overloads for two contrast processes.

It can be seen that the less damaging process (light) has more overloads, and they are distributed with an offset to the beginning of the realization.

When planning an experiment to study the effect of overloads, it is supposed to describe the distribution of overloads along the realization length using a family of beta distributions of a random variable X ~ B (α, β) [8], for which it is true:

$$fx(x) = \frac{1}{B(\alpha,\beta)} x^{\alpha-1}(1-x)^{\beta-1}$$

Where: $\alpha, \beta > 0$ are arbitrary and fixed parameters; $B(\alpha,\beta) = \int_0^1 x^{\alpha-1}(1-x)^{\beta-1}dx$ –beta-function.

For the problem of the distribution of overloads along the length of the realization, the parameters were taken as $\alpha = 1$ and $\beta \sim$ var.

Figure 5 shows two distributions from this family, which can be used to quantitatively describe the distribution of overloads along the length of the realization in the design of an experiment. The distribution with parameters $\alpha = 1$ and $\beta = 5$ describes the situation when the overloads are shifted closer to the beginning of the realization, thus providing a smaller damaging effect (closer to the "light" family, figure 3.). The distribution with parameters $\alpha = 1$ and $\beta = 0.2$ describes the opposite situation- the overloads are moved to end of the realization, and the damage is bigger (closer to the "severe" family, figure 3). To describe the processes, the realization length x and the number of overloads were normalized to unit.

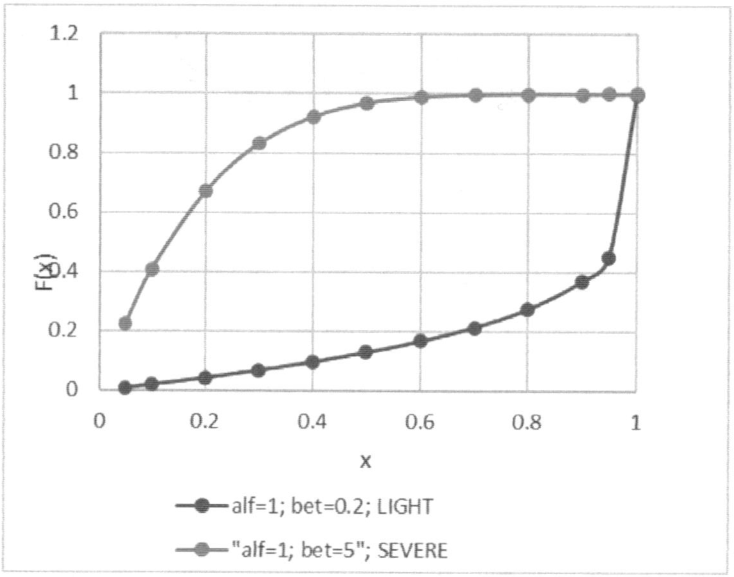

Figure 5. Two distribution from of beta family.

As one can see from the figure 5, the chosen from beta-family distributions, namely, with the parameters $\alpha = 1$ and $\beta \sim 0.2 \ldots 5$ represent well the possible shapes of overloads distribution during the testing.

Conclusions

In the problem of studying the patterns of fatigue crack development, it is desirable to use the theory of optimal experiment planning. Factors selected: asymmetry, frequency, the sequence of peaks. Based on a numerical experiment carried out earlier, proposals were made to formalize the responsible for the sequence of peaks factor.

In the problem of sequence impact on longevity, only one model (namely Willenborg's) was considered. Many other ones exist, and some of them are more advanced and up-to-date (see, for example, [1]). They were not mentioned in this paper.

The question of choosing the testing protocol also matters. Apart from the factors mentioned, one can name the climatic conditions of the testing, the shape of the cycle, the crack length registration procedure, etc. To reasonably decrease the result scatter, the number of tests on one position specimens (in each from 8 cube tops, figure 1) should be taken at least three.

The question remains, what exactly should be taken as a response in planning an experiment: the Paris exponent, the speed on a stable section, or something else. The authors hope to discuss this issue with the conference participants.

Mostly the calculations and modelling, as well as the graphic, were performed in [9].

References

[1] Sunder R 2012 *J. of ASTM Int.* **9(1)** pp 1-32

[2] Romanov A N, Nesterenko G I and Filimonova N I 2018 Damage accumulation under variable loading of cyclically hardening material at the stages of formation and development of cracks *J. of Machinery Manufacture and Reliability* **47(5)** pp 414-19

[3] Lebedinskii S G 2018 Design modeling of propagation of the fatigue cracks in the steel of molded parts of the railway structures *J. of Machinery Manufacture and Reliability* **47(1)** pp 62–66

[4] Gadolina I V *et al* 2020 Studying the crack growth rate variability by applying the Willenborg's model to the Markov's simulated trials *Advances in Intelligent Systems, Computer Science and Digital Economics CSDEIS 2019* **1127**

[5] Makhutov N A *et al* 2020 Modeling a random sequence of extremes of loads for fatigue tests under irregular loading *Vestnik mashinostroenija* **5** pp 13-19

[6] Willenborg J, Engle R H and Wood H A 1971 AFFDL-TM-71-1 FBR WPAFB OH

[7] Adler Ju P 2014 *Introduction into the experiment planning Tutorial* (Moscow: MISIS)

[8] Mosteller F and J Tukey 1977 *Data Analysis and Regression: A Second Course in Statistics* (Addison-Wesley Pub Co)

[9] R Core Team 2020 A language and environment for statistical computing *R Foundation for Statistical Computing Vienna Austria*

Impact of cracks on vibration parameters and lifetime of hydraulic units

Evgeniia Georgievskaia

"Center of design and technological innovation" LLC, Nab. reki Karpovki, d. 5, str. YA, Saint-Petersburg, Russia, 197376

sciencedir@cdti.ru

Abstract. One of the key factors determining the hydraulic unit reliability and lifetime is the current technical state of the runner and, first and foremost, its lack of dangerous cracks. Dynamic load throughout the mode range affects the crack occurrence and development. Blade system crack formation process depends heavily on actual equipment operating modes. Yet, modern hydraulic unit diagnostics systems tend not to record crack formation on its early stages. As a result, routine maintenance often reveals long cracks which require significant financial and time resources to repair. Vibration-based diagnostics systems reveal cracks during operation ineffectively: calculations of the Francis (hydraulic) turbine's runner prove that. High structural rigidity and specific eigenforms don't permit correlating the blade dynamics to measured vibration parameters. Analytic approach based on calculating fatigue strength and fracture mechanics methods could solve the problem mentioned above.

1. Introduction

The important advantage of hydraulic units (HU) is their high maneuverability permitting to start in mere dozens of seconds and reach peak capacity or rapidly change the load. This allows using HU as a main tool in securing sustainability of energy systems as well as compensating power generation instabilities through renewable energy sources (wind, sun, tides, waves, etc.), however, also leading to frequent mode changes, prolonged medium- and low-capacity operation, high number of starts/stops and premature equipment service-life exhaustion [1-2].

This article focuses on high-pressure high-capacity hydraulic units usually equipped with Francis hydraulic turbines boasting high reliability and resistance to cavitation. In Russia Francis turbines of the highest capacity are installed at these hydroelectric power stations: Sayano-Shushenskaya HPP, Krasnoyarsk HPP, Bratsk HPP, Ust-Ilimsk HPP, Bureya HPP and Boguchany HPP.

Runner (RR) is one of the main lifetime-defining parts of the hydraulic unit. Francis turbine runner consists of 11-19 blades, firmly fixed between the lower massive band and the crown. Such blade grid ensures significant structural rigidity of the RR, however, due to geometrical features causes a high level of stress concentration in the connection area between the blade's output edge and the band or crown [3-4].

With the hydraulic unit nearing its peak working efficiency, the maximum blade-system static stress for Francis turbine runner is usually no more than 20-30% of the blade material yield strength, while the dynamic stress not exceeding 10% of the static stress. Such operating mode can't cause crack formation even taking heightened stress concentration into account. However, the blade flow pattern is far from optimal on low/medium partial power modes or when changing modes. This leads to significantly increasing dynamic stress in turbine components and accelerates crack formation [5-7].

As a result, after the HU long-term operation especially including frequent mode-factor changes cracks are often found spreading along the blade-to-crown or band-to-edge welded lines. One usually finds such cracks while checking the unit during routine maintenance [6, 8-9]. At this moment the length of these cracks usually exceeds 300 mm, their opening may be wide and even straight-through. In some cases the uncontrollable crack development leads to complete blade rupture. Blade refurbishment concerning deep and long cracks takes much time and money as well as leads to adverse effects such as high level of welded areas residual stress, redistributed blade loads due to flow part geometry distortion etc.

The problem is no alarming signs of developing defects found in time. Even when using monitoring/diagnostic equipment installed on the HU to perform regular vibration tests, the long-term (years or decades long) crack formation remains undetected. Importantly, modern hydraulic unit monitoring/diagnostic systems analyze data from multiple sensors (sometimes over 200 pcs) installed on static and rotating parts of the unit which is in fact effective for identifying series of operational faults [10]. Unfortunately, cracks developing in the runner are not in the list of the detectable faults. Main reasons for that are explained in this article from mechanical engineer's point of view.

2. Computational models and calculation task

To prove vibration testing systems ineffective for locating emerging cracks in the hydraulic unit runner blade system a computational model of a hydraulic unit rotating part was constructed. It consists of a runner and a hydro-generator, all of them firmly connected to a shaft. To make obtained results more comprehensible the model has the following simplifications:

- RR model uses surface elements while retaining curvature and thickness in each point of the blade without taking actual blades' welded connection area fillet radii into account;
- through cracks in the RR are modelled as ruptures along the connection line between the blade and the crown. These cracks start at the side of the output edge; the length of the rupture increases gradually with the crack's growth;
- the generator model has a simplified view. It allows to retain main mass, dimensions and rigidity parameters along with excluding local effects determined by generator's rotor design features;
- construction fits are assumed to provide tight fit for all the connected elements throughout the mode range;
- presumably, tightening the stud bolts ensures flange joints staying tightly closed under working loads;
- the unit's shaft attachment to guide bearings (upper and lower ones of the generator, the turbine one) and its support by the trust bearing is modelled by anchoring the computational model in relevant cross-sections and directions while maintaining equivalent rigidity rates imitating rigidity of the support.

The first-step calculation task was determining defect-free construction's eigenfrequencies and search for correlations between runner blades' deformation and shaft displacement in areas of actual vibration sensor positioning on a full-scale HU.

The second-step calculations include studying blade crack growth effects on the computational model's eigenfrequencies. Thus, these two options were considered: the growth of a single crack and the simultaneous crack growth on all blades.

In addition to the computational model described above the second step included using refined RR model which shows all the geometry features of the area where blades connect to the crown and the band. The refined model takes RR's cyclical symmetry into account, but doesn't take the unit's shaft and the hydro-generator. The computational model is anchored according to the cross-section relevant to the turbine bearing.

3. Calculation results

Eigenforms and eigenfrequencies on the computational model were defined for 0-160 Hz range. This fully overlaps possible external effect frequency readings concerning the hydraulic unit working in various operating modes. Obtained results are in groups by eigenform type depending on their affecting the possibility of identifying blade system cracks using measurements of unit shaft's vibratory displacements.

Figure 1 shows main reviewed computational model eigenform types. They present different variants of correlation between runner deformations and radial shaft displacements close to supports – places having vibration sensors installed on a full-scale hydraulic unit:

a) bending vibrations of the whole unit itself when there is a correlation between runner displacements (deformations) and radial displacement of the shaft detected by shaft-runout sensors close to bearings;

b) torsional vibrations without shaft displacements in a radial direction during the runner twist; registered vibrational parameters don't usually include the shaft twist; accordingly, the vibration monitoring system doesn't identify such vibrations;

c) bending vibrations of the runner without HU shaft displacement close to support components; vibratory shaft displacement is absent despite the present runner deformation; the vibration monitoring system doesn't form a warning signal even if resonant vibrations of such type are present.

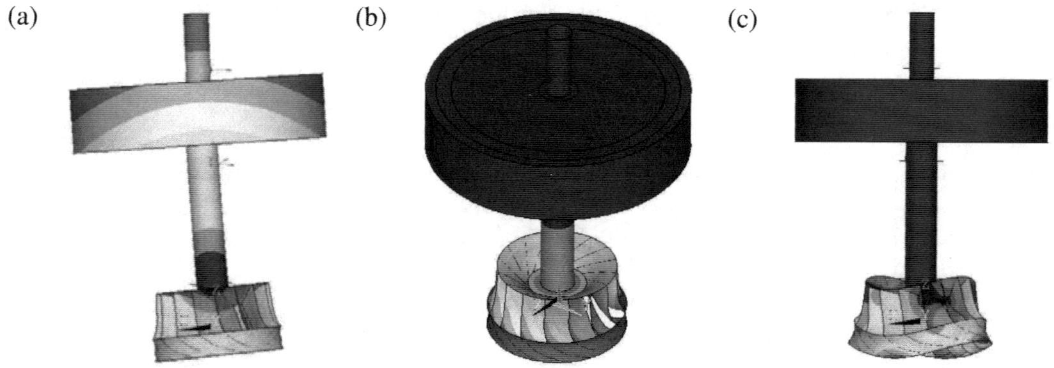

Figure 1. Main eigenform types for a defect-free construction: (a) – bending vibrations of the unit's shaft, (b) – torsional vibrations of the runner without radial shaft displacement, (c) – bending vibrations of the runner without HU shaft deformation.

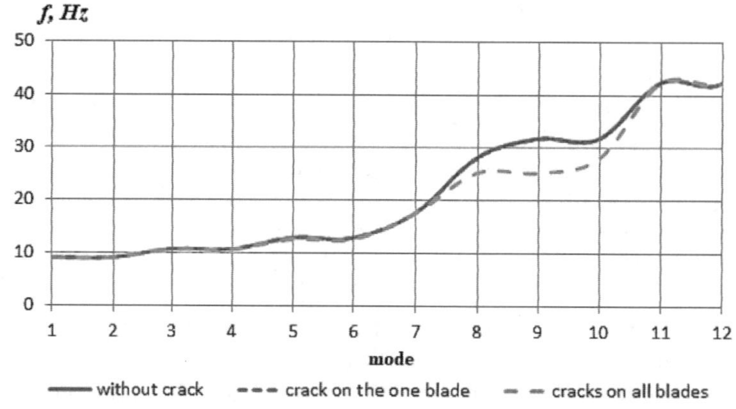

Figure 2. The effect of a single or multiple cracks on unit's eigenfrequency.

Figure 2 shows comparative unit's eigenfrequencies range calculation results for three cases in general: no cracks on RR blades, a crack on one RR blade, and cracks of equal length on all RR blades. Crack length is considered to be 10% of the length of the joint between the blade and the crown. The x-axis represents the eigenform number: first 12 values given fall within the range of 0-50 Hz. The y-axis sets eigenfrequency values of the computational model. Vibration-based diagnostics system is able to identify such values under operating conditions of the unit at a hydroelectric power station (for example, based on analyzing recorded vibration displacement of the turbine bearing).

Figure 3 shows selected results for the refined model: the first eigenfrequency of the blade without cracks and the blade with a through crack of different lengths L: $L=0.14L_0$ and $L=0.25L_0$, where L_0 is the total length of the line where the blade is welded to the RR crown.

Blade eigenfrequency shift (Δf, %) is calculated using relative numbers: the remainder of the non-cracked blade eigenfrequency subtracted by the eigenfrequency of the blade with a crack of L length, divided by the defect-free construction's eigenfrequency. The stress-strain state correlating to the runner eigenform was defined using the first eigenfrequency. The effect of eigenfrequency shift towards the change in the vibration parameters available for measuring on a full-scale hydraulic unit non-stop operation was investigated. There have also been similar numeric tests for higher eigenfrequencies.

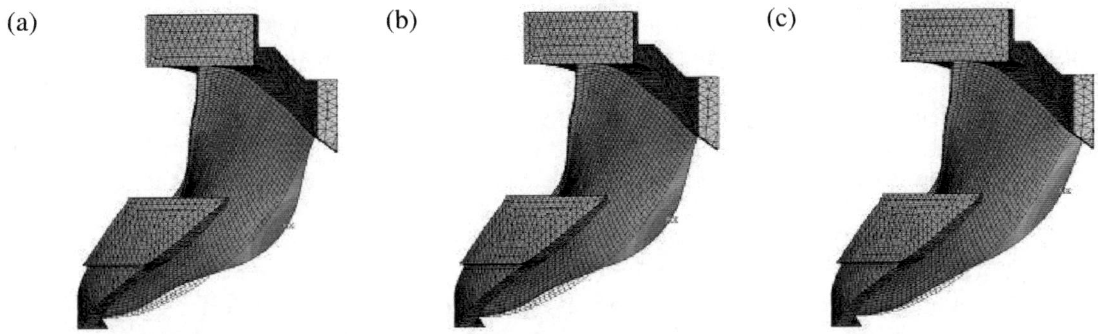

Figure 3. The first form of blade eigenfrequencies with different crack length: (a) L=0 mm, Δf = 0, (b) L=0.14L0, Δf=4.4%, (c) L=0.25L0, Δf=17.8 %.

Figure 4 shows results of relative crack length L (percentage of total joint between the blade and the crown) effect on eigenfrequency value (f). Eigenfrequencies were calculated using the refined model. The report presents data for the initial three eigenvalues of a single blade with a growing crack. Modelled through-cracking began from the output edge and was spreading along the attachment line between the blade and the crown.

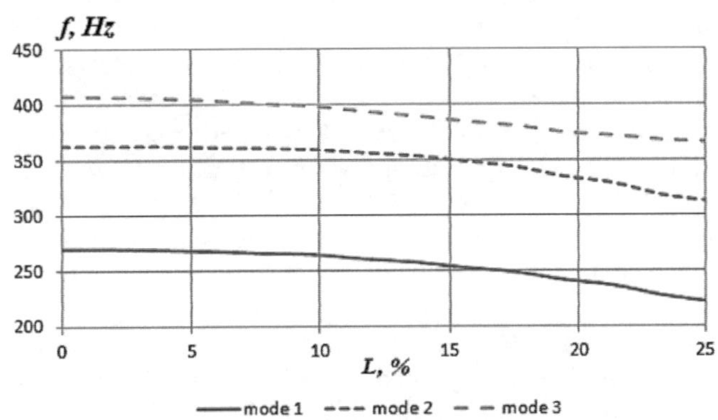

Figure 4. Relation between the eigenfrequency and the blade crack length.

4. Discussion of results

The results presented in figure 1 prove identifying torsional vibrations and runner bending without shaft bending using vibration sensors installed close to the supports is impossible. The same applies to axial eigenform of the runner and any hybrid forms which possess components presented in figure 1(b) and 1(c). Therefore, in these cases blade dynamic stress level cannot be defined which leads to inability of estimating the remaining lifetime of a HU using vibration monitoring data on an operating HU.

Blade crack development leads to decreasing construction eigenfrequencies and support element reaction change. However, as conducted calculations show (see figures 2-4), high construction rigidity of a Francis turbine runner does not allow identifying these changes at an early stage. For example, changing of the first eigenfrequency of a separate blade for 5% corresponds to a through crack with a length close to 15% of total joint between the blade and the crown (see figure 4). At the same time, unit shaft eigenforms and eigenfrequencies change by less than 0,5% (see figure 2). This is practically impossible to identify under actual operating conditions. Although the development of a long, even though, crack leads to changing the blade's eigenform (maximum deformation gradually shifts from the middle of the output edge closer to the crown in step with crack growth), this almost doesn't affect crown or runner band form changes, not to mention the shaft or hydraulic unit supports. That's why such changes are difficult to register using vibration monitoring data analysis.

Calculated testing has shown that separate element identifiable changes throughout the shaft eigenform monitoring system range will occur only when cracks in all the runner blades longer than 10% of total joint length are present (see figure 2). Obviously, such situation is unlikely to occur as blade cracks form unevenly and grow with different speeds. This is primarily related to the lack of RR geometry perfection and to the presence of disk eigenforms (see figure 1(c)) and the fact that significant dynamic stress distribution unevenness throughout the blades accompanies them.

5. Conclusion

Calculation results in the article show the following: high construction rigidity and specific eigenform presence do not allow to correlate blade dynamic stress levels (also responsible for emergence and developing of cracks) to measured hydraulic unit vibration monitoring/diagnostics systems. This makes detecting cracks in time on an operating hydraulic unit difficult. Solving the problem is possible through approaches which implement analytical method based on calculating fatigue strength and crack growth speed using fracture mechanics approaches [11].

References

[1] Trivedi C, Gandhi B and Michel C J 2013 Effect of transients on Francis turbine runner life: a review *J. of Hydraulic Research* **51(2)** pp 112-32 doi: 10.1080/00221686.2012.732971

[2] Monette C, Marmont H, Chamberland-Lauzon J, Skagerstrand A, Coutu A and Carlevi J 2016 Cost of enlarged operating zone for an existing Francis runner *IOP Conf. Series: Earth and Environmental Science* **49** 072018 doi:10.1088/1755-1315/49/7/072018

[3] Flores M, Urquiza G and Rodríguez J M A 2012 Fatigue Analysis of a Hydraulic Francis Turbine Runner *World J. of Mechanics* **2** 28-34 doi:10.4236/wjm.2012.21004

[4] Huang X, Oram C and Sick M 2014 Static and dynamic stress analyses of the prototype high head Francis runner based on site measurement *IOP Conf. Series: Earth and Environmental Science* **22(3)** 032052 doi:10.1088/1755-1315/22/3/032052

[5] Unterluggauer J, Doujak E and Bauer C 2019 Fatigue analysis of a prototype Francis turbine based on strain gauge measurements *J. WasserWirtschaft* **09(S1)** 66-71 doi: 10.1007/s35147-019-0238-92

[6] Liu X, Luo Y and Wang Z 2016 A review on fatigue damage mechanism in hydroturbines *J. Renewable and Sustainable Energy Reviews* **54** 1–14 doi: 10.1016/j.rser.2015.09.025

[7] Seidel U, Mende C, Hübner B, Weber W and Otto A 2014 Dynamic loads in Francis runners and their impact on fatigue life *IOP Conf. Series: Earth and Environmental Science* **22(3)**

32054 doi:10.1088/1755-1315/22/3/032054

[8] Nennemann B, Monette C and Chamberland-Lauzon J 2016 Hydrodynamic damping and stiffness prediction in Francis turbine runners using CFD *IOP Conf. Series: Earth and Environmental Science* **49(7)** 072006 doi:10.1088/1755-1315/49/7/072006

[9] Boukani H Habibzadeh, Viens M, Tahan SA and Gagnon M 2014 On the performance of nondestructive testing methods in the hydroelectric turbine industry *IOP Conf. Series: Earth and Environmental Science* **22(1)** 12018 doi:10.1088/1755-1315/22/1/012018

[10] Andritz Hydro: DIA TECH Monitoring and Diagnosis System. Available online http://www.andritz.com/hy-diatech-monitoring-diagnosis.pdf (accessed on 1 February, 2017)

[11] Georgievskaia E 2019 Hydraulic turbines lifetime in terms of fracture mechanics *J. Engineering Failure Analysis* **105** 1296-305

Acknowledgments

This research is being carried out within the framework of the Skolkovo project.

This research was funded by the Foundation for Assistance to Small Innovative Enterprises in Science and Technology (FASIE), Russia, grant number 3288ГС1/55639 and the Skolkovo Innovation Center, grant number 40113/07002/0329-2020.

Ultrasonic detection of manufacturing defects in multilayer composite structures

A M Kokurov[1,2] **and D E Subbotin**[2]

[1] Mechanical Engineering Research Institute of the Russian Academy of Sciences, 4 Maly Kharitonyevsky Pereulok, Moscow, 101990, Russia
[2] PJSC "Tupolev", 17 Academician Tupolev Embankment, Moscow, 105005, Russia

and-kokurov@yandex.ru

Abstract. The authors have considered practical aspects of the use of ultrasonic method for identification of foreign objects in the structure of laminated composite. The paper presents the results of laboratory investigations aimed at determination of shape and dimensions of artificially created defects and the analysis of specific aspects of ultrasonic detection of fragments of prepreg protective coating.

1. Introduction

Nowadays, composite materials are actively used in production of advanced high-strength structural elements. Their extensive use is directly related to the unique specificity of the structure of this material which enables the manufacture of designs with predictable mechanical properties [1, 2]. Special semi-finished products – prepregs – are used in the process of manufacture of laminated composite structural elements. They consist of fibrous reinforcing filler and the binding substance. Carbon fibers are most commonly used in technical objects, where the binding substance is the epoxy resin. Autoclaving is most commonly used in manufacturing of finished designs [3].

The process of composite package formation calls for strict adherence to optimal technological mode at each stage of manufacture. Non-compliance with manufacturing process may provide the generation of various defects in the structure of the material [4]. The presence of such defects may significantly reduce the strength of the manufactured structure [5].

The most common material structural defects are the non-homogeneity of various shapes and dimensions [6]. The experience of the use of these materials demonstrates that one of the main causes of occurrence of defects in the laminated material structure is the presence of residues of transportation PE film (prepreg protective coating). In the process of composite package formation, the joint melting of the binding substance and the film fragments takes place. It significantly complicates the identification of these material structure damage by the present-day non-destructive methods [7-13]. Therefore, the solution of the task of manufacturing defects identification in composite structures is important and essential.

This paper considers the practical aspects of the use of ultrasonic method for detection of process defects and structural non-homogeneity of material, as well as foreign inclusions of various dimensions and shapes in polymer laminated composites.

2. Control means

The physical principle of the considered non-destructive testing method is based on the fact that hard materials are good sound waves conductors. Reaching the boundaries of surfaces or internal defects (cracks, pores, various inclusions, etc.), the waves reflect from them [14-16]. Ultrasonic attenuation in polymer composite materials has high frequency dependence [17]. Therefore, in the process on non-destructive testing of such materials it is necessary to reduce the ultrasonic range frequencies [18]. In particular, ultrasonic testing of polymer binding-based composite materials is possible at the frequencies below 2 MHz.

In the process of laboratory testing, the authors used the inspection device UD4-T and various types of direct combined P111-series transducers. The direct combined transducers are used for longitudinal waves testing of various articles. The effective frequency of the used transducers varied in the range from 1 to 2 MHz. The average damping degree provided the optimal ratio between the pulse amplitude and the resolving power. Various transducers were used to improve the accuracy and reliability of the results obtained. The size of piezoelectric cells in them varied from 6 to 20 mm.

3. Study objects

The objects of the demonstration tests were rectangular plates with artificial defects. The plates were manufactured of a unidirectional prepreg with high-strength carbon fiber and polymer binding. The composite package consisted of 36 0.113 mm thick mono-layers and had the following monolayer orientation angles: +45, 0, -45, 90, 0, 0, 0, +45, 0, -45, 0, 0, 0, 0, +45, 90, 0, 0, 0, 0, 90, +45, 0, 0, 0, 0, -45, 0, +45, 0, 0, 0, 90, -45, 0, +45 (degrees). All composite plates were of the same dimensions 100x150 mm.

In the process of composite package forming, the defect simulators – the fluoroplastic film fragments – were placed between the layers. The film fragments differed in shape and dimensions but has similar thickness of 0.12 mm. The simulators were placed in the center of the sample at various distances from its surface (figure 1). The total of four samples were manufactures with the following variations of defect simulator shape and dimensions:

1. rectangular shape, simulator dimensions: 55×83 mm, the distance from the surface to the defect is 0.678 mm (between the layers 6 and 7);

2. round shape, simulator diameter is ø 31 mm, the distance from the surface is 0.678 mm;

3. round shape, simulator diameter is ø 62 mm, the distance from the surface to the defect is 0.678 mm;

4. round shape, simulator diameter is ø 62 mm the distance from the surface to the defect is 1.356 mm (between the layers 12 and 13).

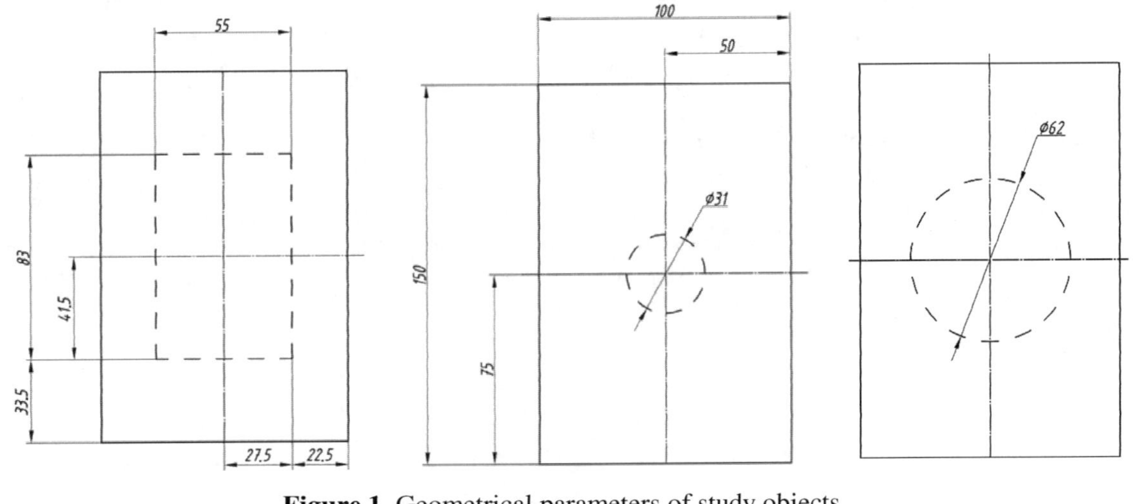

Figure 1. Geometrical parameters of study objects.

4. Diagnostic methods and results

The diagnostics of each of the samples consisted of several steps. At the first step, in the process of scanning, the sensor moved horizontally along the long side of the sample. At the second and the subsequent steps, the scanning was performed along the short size (vertically) and along the diagonals of the sample correspondingly. At each scanning the readings and the sensor positions, necessary for the further processing of the results, were recorded. At detection of the defect boundary, a graphic marking was made on the surface of the sample, at the point of current location of the transducer. Upon completion of the test, the contour of the detected defect was restored from the marking. The test patterns obtained are presented in figure 2.

Sample 1

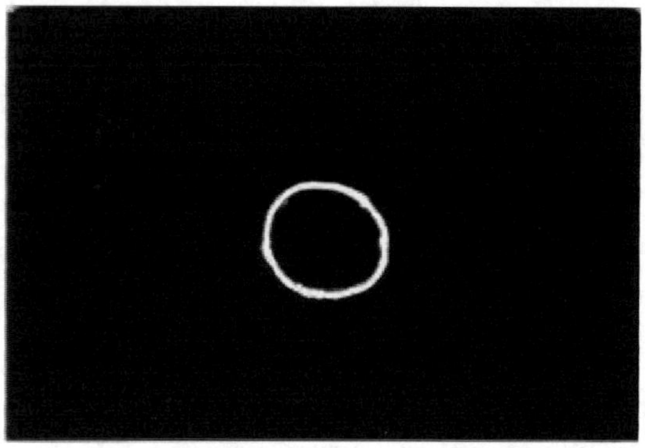

Sample 2

Sample 3

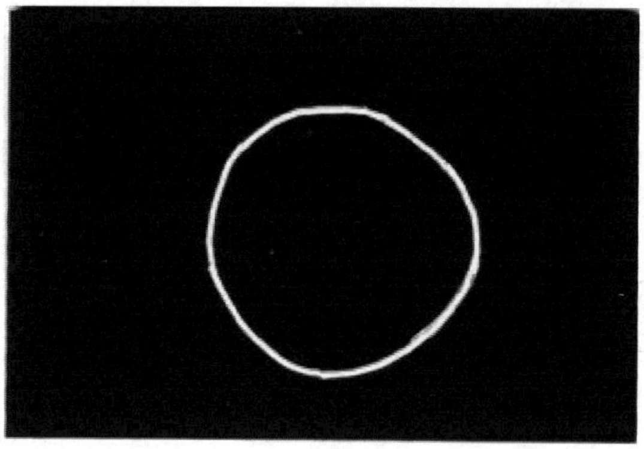

Sample 4

Figure 2. Scanning results.

The digital twin of the sample was used to assess the error of detection of geometry and dimensions of the defect. The twin was the graphical presentation of the object obtained by digitization of the surface of the sample with defect contour marked. To calculate the defect areas, a software providing the calibration by the known dimension with the following calculation of geometrical parameters was used. The obtained scale factors were used to recalculate the dimensions of the defect area into the linear dimensions X (pixels), Y(pixels) → X(mm), Y(mm). The calculation results are presented in the table 1.

Table 1. Defect dimensions measured.

Sample number	Flaw dimensions		Error	
	Nominal, mm	Measured,mm	Absolute, mm	Relative, %
1	55	52	3	5.4
	83	78	5	6
2	31	29	2	6.4
3	62	58	4	6.4
4	62	64	2	3.2

Analyzing the graphical and numerical test results, it can be concluded that all the defect simulators were detected by the ultrasonic method. The accuracy of detection of their geometrical parameters is high. Besides, it is possible to note that with the increase of the distance from the surface, and, correspondingly, the sensor to the defect, the accuracy of detection of its dimensions increases.

5. Conclusion

Non-destructive ultrasonic testing method can be efficiently used for detection of structural defects having the similar physical and chemical properties as the binding material. The use of advanced hard- and software for testing results processing provides high accuracy of detection of defect dimensions and location.

The obtained laboratory tests results make it possible to recommend the obtained method and technical solutions for diagnostics of laminated polymer composite structures by non-destructive ultrasonic testing method.

References

[1] Zhu L, Li N and Childs P R N 2018 Light-weighting in aerospace component and system design *Propulsion and Power Research* vol **7** issue 2 pp 103-19

[2] Boyer R R, Cotton J D, Mohaghegh M and Schafrik R.E 2015 Materials considerations for aerospace applications *Materials & Engineering: Propelling Innovation* vol **40** issue 12 pp 1055-66

[3] Park S Y, Choi C H, Choi W J *et al* 2019 A comparison of the properties of carbon fiber epoxy composites produced by non-autoclave with vacuum bag only prepreg and autoclave process. *Applied Composite Materials* 26 pp 187–204

[4] Talreja R 2015 Manufacturing defects in composites and their effects on performance *Polymer Composites in the Aerospace Industry* pp 99-113

[5] Elhajjar Rani, Grant Peter N and Ashforth Cindy 2018 *Composite Structures: Effects of Defects.* (New York: John Wiley & Sons Ltd)

[6] Murashov V V and Rumyantsev A F 2006 Types of production and operational defects of structures made of polymer composite materials *VI Scientific Conf. on Hydroaviation "Gidroaviasalon-2006"* pp 147-53

[7] Wronkowicz-Katunin A, Mihaylov G, Dragan K *et al* 2019 *J. of Nondestruct Eval* **38** 82

[8] Kanevsky I N 2007 *Non-destructive control methods* ed I N Kanevsky and E N Salnikov (Vladivostok: Publishing house of the Far Eastern State Technical University)

[9] Zhao Q, Dan X, Sun F, Wang Y, Wu S and Yang L 2018 Digital shearography for NDT: Phase measurement technique and recent developments *Applied Sciences* **8(12)**

[10] Malushin D S 2019 Multiparameter control of multilayer structures by pulsed eddy-current techniques *Technical Phys. Letters* vol **45(8)** pp. 757-60

[11] Shkatov P N *et al* 2019 Study of interaction of eddy current probes and delamination in multidirectional CFRP *Materials Science Forum* vol **970** pp 305-13

[12] *Introduction to phased array ultrasonic technology applications*: *R/D tech guideline* 2004 (Quebec: R/D Tech inc)

[13] Boychuk A S, Stepanov A V, Kosarina E I and Generalov A S 2013 Application of ultrasonic phased array technology in non-destructive testing of aircraft parts and structures made of PCM *Aviation materials and technologies* **2(27)** pp 41–46

[14] D'orazio T, Leo M, Distante A, Guargnella C, Pianese V and Gavaccini G 2008 Automatic ultrasonic inspection for internal defect detection in composite materials *NDT & E Int.* **41(2)** pp 145–54

[15] Benammar A, Drai R and Guessoum 2008 A Detection of delamination defects in CFRP materials using ultrasonic signal processing *Ultrasonics* **48(8)** pp 731-38

[16] Kappatos V *et al* 2017 Theoretical assessment of different ultrasonic configurations for delamination defects detection in composite components *Procedia CIRP* vol **59** pp 29-34

[17] Williams J H, Nayeb-Hashemi H and Lee S S 1980 Ultrasonic attenuation and velocity in AS/3501-6 graphite fiber composite *J. Nondestructive Evaluation* **1** pp 137–48

[18] Honarvar F, Sheikhzadeh H, Moles M and Sinclair A N 2004 Improving the time-resolution and signal-to-noise ratio of ultrasonic NDE signals *Ultrasonics* **41(9)** pp 755–63

Acknowledgments
This work was supported by the Russian Science Foundation (project № 20-19-00769).

Structural quality criteria for a surface layer formed by gas pressure nitriding of austenitic steels

L I Kuksenova[1], M S Alekseeva[1,2], M A Gress[3] and D A Kozlov[1]

[1] Mechanical Engineering Research Institute of the Russian Academy of Sciences, 4 Small Kharitonyevsky lane, Moscow, 101000, Russia
[2] Federal state unitary enterprise all-russian scientific research institute of aviation materials, State scientific center of the Russian Federation,17 Radio str, Moscow, 105005, Russia
[3] Bauman Moscow State Technical University, 5 bil. 1 Second Baumanskaya str, Moscow, 105005, Russia

lkukc@mail.ru

Abstract. The tribotechnical efficiency of gas pressure nitriding of austenitic class steels is shown. The characteristics of the structure and operational properties (contact endurance and wear resistance) are given in comparison with cemented steels.

1. Introduction

Nitriding along with cementation are the main methods of surface hardening of the most critical machine parts. An important advantage of nitriding in comparison with cementation is the provision of a higher extreme pressure resistance of steel, cyclic strength, and corrosion resistance. The advantage of nitriding is a slight change in the volume of parts, which can be taken into account in their manufacture. Grinding volumes are incomparably smaller than with cementation. The noted advantages explain the continuous growth in the use of nitriding in various engineering industries. Along with this, gas or ion nitriding has significant drawbacks, the main of which are a long saturation time and a limited (up to 0.5 mm) thickness of the diffusion layer.

The prospect of eliminating these shortcomings and a technological breakthrough in the field of increasing the load capacity of steel parts reinforced with nitrogen is associated with the development and application of gas-pressure nitriding [1]. Gas-pressure nitriding is a process that implements the simultaneous impact on the surface of steel of high nitrogen gas pressure and temperature, which provide a high rate of nitrogen supply to the saturated surface and a high rate of its diffusion removal into the metal depth. These features make it possible to form diffusion layers of large thickness, commensurable with the thickness of the cemented layer in a relatively short time - 3-5 hours.

Austenitic structural steels have a valuable combination of mechanical and corrosive properties. Defect of these steels is the low resistance to fatigue and wear, which limits the scope of their application. The main way to increase the operational characteristics of austenitic steels is nitriding. At the same time, high-chromium steels of the austenitic class are hardly nitrided. Therefore, the task arose of finding ways to improve nitriding and intensification of the process of diffusion saturation of austenitic steels. In this paper, we consider the use of gas pressure nitriding and the influence of technological process factors on the structural characteristics and operational properties of austenitic steels is shown.

2. Materials and research methods

Experimental samples of high-chromium austenitic steels 12Kh18N10T, 08Kh16G15N5MAF and high-manganese steel 40G14N8Kh3B1 were subjected to gas-pressure nitriding. The pressure of the gaseous medium was varied from 30 to 150 MPa, the process temperature was in the range of 950 - 1150°C, the saturation time was 3 hours. Bench tests to assess the resistance of the nitrided layer of contact fatigue were performed on a Sh-17 two-pin roller machine of the IMASH RAS design, which provided tests according to the rolling pattern with 6% slip and imitated the operation of the nitrided layer of gears [2]. Tests for wear resistance were carried out on a Skoda-Savin machine with a load of 150N. The structure of the nitrided layer was analyzed by x-ray and electron microscopy methods.

3. Research results and discussion

The effect of process pressure on the microhardness distribution of 12Kh18N10T steel is shown in figure 1.

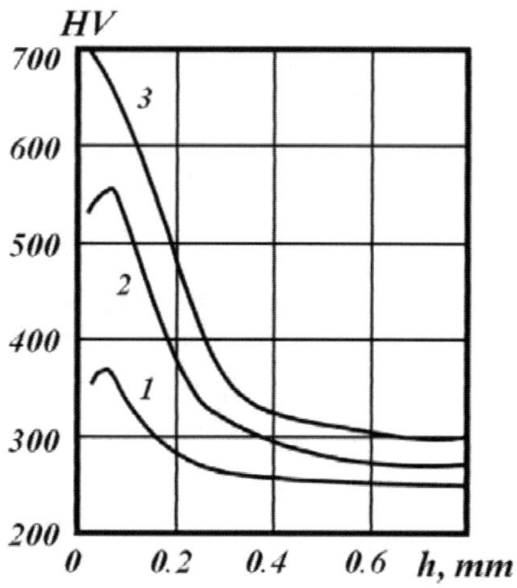

Figure 1. Microhardness distribution over the thickness of the nitrided layer of 12Kh18N10T steel saturated at 950°C, 3 h and pressure: 1 - 30 MPa; 2 - 100 MPa; 3 - 150 MPa.

It is seen that there is a flowing distribution of hardness over the thickness of the layer. A noticeable increase in surface hardening is observed with increasing nitrogen pressure in the working chamber. This dependence indicates that an increase in pressure increases the saturating ability of the technological atmosphere and, as a result, a more intensive supply of nitrogen to the surface layer. In the presence of a high chromium content in steel, the incoming nitrogen binds to highly hard chromium nitrides. The density of nitride particles, their size, and their relation to the matrix lattice largely determine the level of surface hardening. Meanwhile, the outflow of atomic nitrogen to the growing nitride particles inhibits the growth of the diffusion layer, the thickness of which turned out to be practically independent of pressure at a temperature of 950°C and amounted to about 0.5 mm. According to transmission electron microscopy, the main components of the structure of the nitrided layer are martensite, austenite, and CrN nitrides, the volume fraction of which is not less than 20%. Nitride particles are densely located in the martensitic matrix, have an equiaxial shape and a size of ~ 10 15 nm, figure 2, and orientation in the region of martensite plates, which illustrates the dark-field image in the (111) CrN reflection. The structure of the studied steels 40G14N8Kh3MF1 and 08Kh16G15N5MAU is similar to the structure of the layer of steel 12Kh18N10T considered, while the nitride formation intensity increases in proportion to the concentration of chromium in the steel.

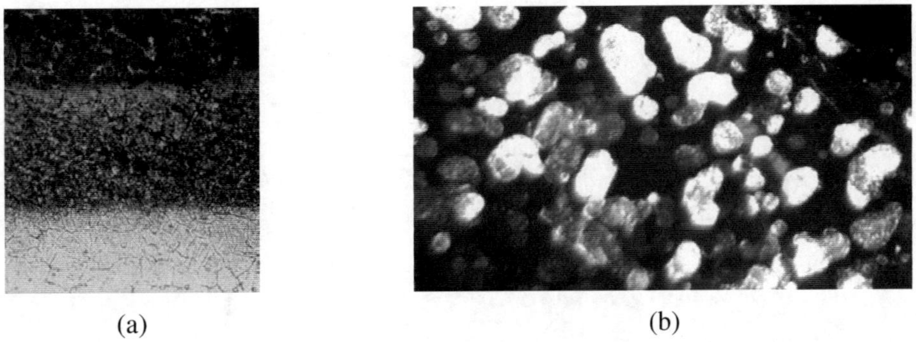

Figure 2. The structure of the nitrided layer of steel 12Kh18N10T after diffusion saturation at 150 MPa, 1050°C, 3 h: (a) - microstructure of the layer (transverse section), x250; (b) - electron microscopic dark-field image in the reflection of nitride CrN, x50 000.

The effect of nitriding temperature is illustrated by the data in table 1.

Table 1. Characteristics of the diffusion layer.

Steel grade	Pressure, MPa	t, °C	Layer thickness, mm	Hardness	Phase comp.[a]
12Kh18N10T	30	950	0.35	380	M, A, CrN
	100	950	0.50	550	
	150	950	0.60	730	
	150	1050	0.90	790	
40GL14N8Kh3MF1	150	950	0.35	450	M, A, CrN, VN
	150	1050	0.85	540	
08Kh16G15N5MAU	150	950	1.10	590	M, A, CrN, AlN
	150	1050	1.60	680	

[a] M-martensite, A-austenite.

It is shown that an increase in the pressure and temperature of the process allows reaching the thickness of the hardened layer of 1.5 mm, which significantly (3-4 times) exceeds the capabilities of traditional processes of gas and ion nitriding. The higher the chromium content, the greater the surface hardening. This result confirms the position that chromium nitrides make the main contribution to austenite hardening.

In this regard, the results of bench tests for contact endurance and wear resistance were analyzed from the standpoint of structural changes. Examples of test results in the form of contact fatigue curves are shown in figure 3. It can be seen that the greatest contact endurance limit is ensured by gas pressure nitriding at 150 MPa, 1000°C, with quenching and subsequent tempering at a temperature of 500 °C, curve 1.

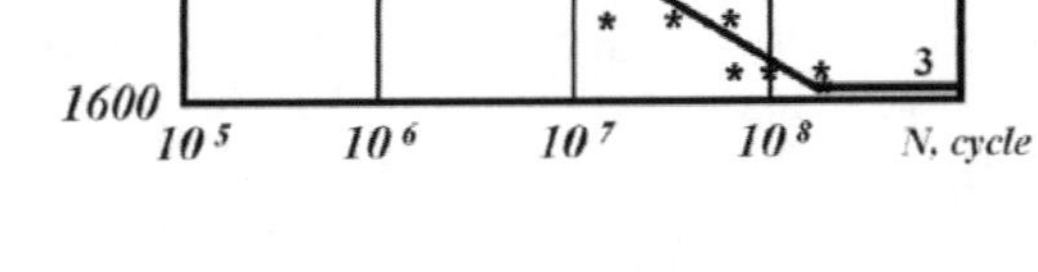

Figure 3. Contact strength curves of steel: 1 - gas pressure nitriding; 2 - cementation; 3 - gas nitriding

Wear resistance was studied on steels martensitic 16Kh2N3MFBAY-Sh and austenitic 12Kh18N10T classes. Similar data were obtained for steels and unified laws obtained.

Table 2. Relative wear resistance.

Preasure, MPa	Temperature, °C	Tempering temperature, °C	Relative wear resistance
30	950	-	1.17
100	950	-	1.28
150	950	-	1.39
150	1050	-	1.71
150	1100	-	1.76
150	1000	-	2.13
150	1000	-	2.19
Gas cementation, 930°C, 8h			1.56
Gas nitriding, 540°C, 48h			2.83

An analysis of the results of metallophysical studies showed that the effect of the sizes of nitrides on contact endurance and wear resistance is associated with the level of microdeformation of the crystal lattice of the matrix, which depends on the coherent or incoherent release of nitrides [3]. Based on the analysis of the broadening of the martensite lines, it was found that with coherent nitrides the magnitude of microdistortions is about 10.3×10^{-3} with incoherent particles it is 2.5 times smaller. With incoherent precipitates, the density of the nitrides of the particles is lower, the fragility is less, and as a result, the fatigue resistance is higher. A decrease in the magnitude of microdeformation leads to an increase in the plasticity in the microvolumes of the layer, which contributes to the relaxation of local microstresses and to an increase in resistance to the nucleation of a fatigue crack and generally to an increase in the operational characteristics of steel.

4. Conclusion

The advantages of gas pressure nitriding over existing processes of alloying the surface of steel with nitrogen are established. Particularly great is the advantage of this process when hardening austenitic steels belong to hard nitriding. For these steels, gas pressure nitriding is the only process that makes it possible to obtain layers of high hardness and wear resistance with a thickness exceeding 1 mm. The characteristics of diffusion layers — their structural state, saturation, phase composition, and, as a result, properties are controlled by a small group of technological factors — pressure, temperature and time parameters.

References

[1] Gerasimov S A, Zhikharev A V, Berezina E V, Zubarev G I and Pryanihnikov V A 2004 New ideas on the mechanism of formation of the structure of nitrided steels *Metal Science and Heat Treatment* (Moscow) **1** pp 13-17

[2] Eliseev Yu S, Krymov V V, Nezhurin I P, Novikov V S and Ryzhov N M 2001 Production of gear wheels of gas turbine engines (Moscow: Higher school) p 493

[3] Gerasimov S A, Kuksenova L I, Lapteva V G, Ospennikova O G, Alekseeva M S and Gromov V I 2019 *Surface engineering and operational properties of nitrided structural steels* ed E N Kablova (Moscow: VIAM) p 600

Relation of threshold levels fatigue crack development in low-alloy steel under harmonic and operational loading

S Lebedinsky and O Naumov

Mechanical Engineering Research Institute of the Russian Academy of Sciences, 4 Maly Kharitonyevskiy Pereulok, Moscow, 101990, Russia

SLebedinski@yandex.ru

Abstract. The results of experimental studies of the patterns fatigue crack development in samples made of 20GFL steel cut from the cargo over-spring beam are analyzed. The relatio of the threshold stress intensity coefficient Kth, determined from the kinetic diagram of fatigue failure, and the parameters of the threshold level and the operational loading process is estimated.

Introduction

The task of improving estimation accuracy of the of railway structures [1, 2] resource is of paramount importance for cast parts. It is known that in the process of their manufacture, occurrence of casting defects is inevitable, resulting in formation of stress concentrators further leading to the development of fatigue cracks. This way, the resource of a freight car bogie is largely determined by the duration of development of fatigue cracks to critical values. As studies [3-5] show the main factors affecting the development of cracks at low, pre-threshold loading levels are the effect of cyclic hysteresis of stresses and deformations at the apex of a crack, and their interaction with residual stresses. At the same time, with the reduction of fatigue crack growth rate, the impact of the environment on the fracture process rises significantly. This indicates the complexity of forecasting of damage accumulation processes when reducing operational loads, especially considering their stochastic nature.

As a rule, calculations of fatigue cracks development, influenced by operational loads are carried out using parameters of the kinetic diagram of the fatigue failure of the material, which are obtained experimentally under harmonic loading. And since one of the most important parameters is the threshold level of the stress intensity factor Kth, it is important to establish the relation of this value with the corresponding average parameters of the operational loading process. Then, the Kth, determined on the fatigue failure kinetic diagram (FFKD) of 20GFL steel [6] and the parameters of stress intensity factor (SIF) variation process, corresponding to the condition of the crack arrest on the same steel grade specimen were compared, when simulating the operational loading process. During the above comparison procedure, analysis of the work [7] is concurrently carried on. This paper presents a method for determining the crack development rate variation pattern under operational loading reduction. The method consists in controlling the loading level reduction in accordance with specimen stiffness decrease (increase in compliance) while the crack grows. During the first stage of experiment, the loading block recorded in operation is reproduced, and the corresponding process of crack edges opening is recorded (figure1). In the second stage, this recording is activated on the test bench as a control signal for reproducing the operational load block.

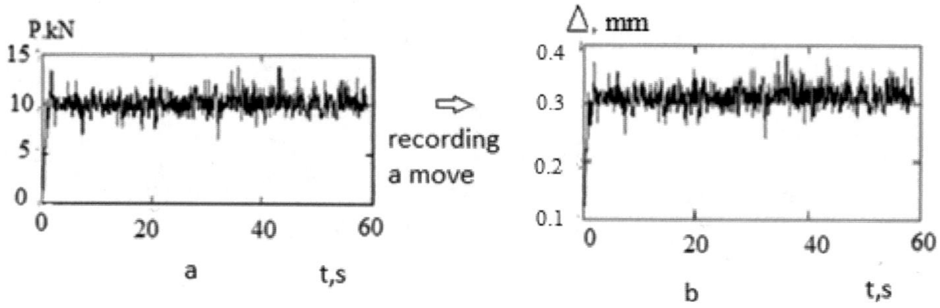

Figure1. Sequence of formation of the control signal for hard (on the opening) loading at the second stage, (a) - the loading process, (b) - corresponding process of crack edges displacement.

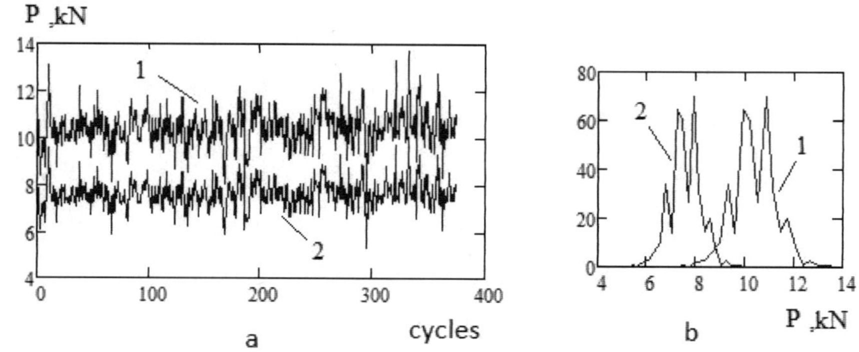

Figure 2. Reduction in the operational level of loading with an increase in the specimen compliance from crack growth from the 1st level to the 2nd - (a), while maintaining the similarity, (b) - the corresponding shift in the distribution of extremes of processes 1 and 2.

With the development of the crack and reduction of the specimen stiffness, the loading level drops. Analysis of the regularity of changes in the parameters of the simulated loading process with concurrent crack growth and its level corresponding reduction, shows that the ratio of maximum and minimum values of cycles (extremes) in the loading block remains constant. Process of reduction of average, general loading level is under way. This allows us to consider the loading process with crack growth similar to changing processes, with a similarity coefficient matching the reduction level (figure 2).

Applying extrapolation to the results of such tests, a method for calculating and experimentally predicting the rate of development of fatigue cracks in the studied steels at a low level of operational loading, while maintaining the constant ratio between the minimum and maximum values of the cyclic variable loading, was developed [7].

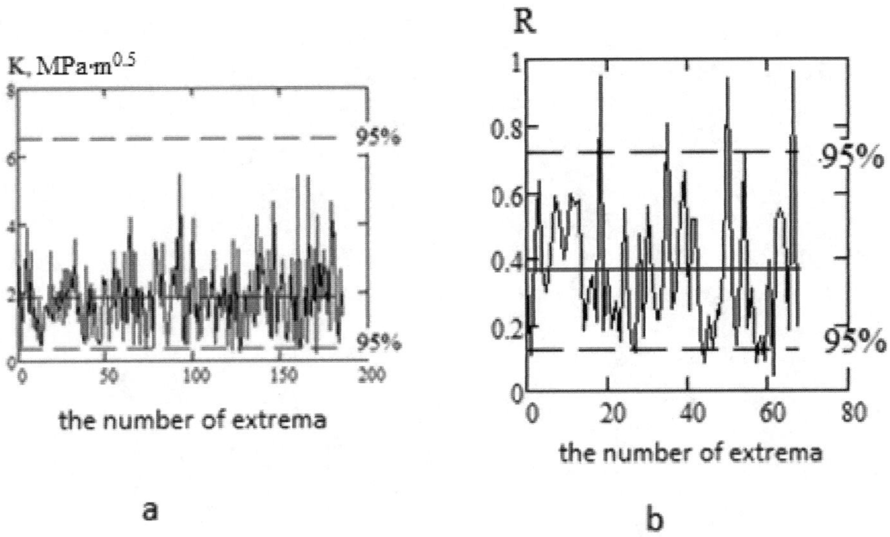

Figure 3. (a) - block of operational loading in the SIF units that meets the conditions for cracks arrest in 20GFL steel [7] and 95% confidence limits of the process; (b) - process of the skewness coefficient R change in accordance with the SIF change (figure 1). The average value of R=0,364 and 95% confidence limits of the SIF.

Test results

In the experiment [7], the relationship of fatigue crack development rate reduction and the number of repeating blocks of operational loading, and the corresponding load decrease under the fixed process of crack edges opening. Extrapolation of the load reduction value ΔP to the level of the zero velocity value (the crack arrest condition) was performed using regression dependencies. The similarity coefficient predicted for the zero speed level allows us to build a model of the threshold loading process, expressed as a cyclic SIF change. The process obtained from the results of work [7] for a compact specimen of C(T) type is shown in figure 3 (a). The following parameters of the resulting process are determined. The average SIF value of the process is 1.978 MPa·$m^{0.5}$, in the range of 0.2036 - 5.458 MPa·$m^{0.5}$, with a standard deviation of 1.235. Figure 3 (a) shows 95% confidence limits of the SIF (dotted lines), assuming a lognormal distribution. Similar calculations were carried out for the sequence of values of the skewness coefficients R, defined as the ratio of successive SIF extremes (figure 3 (b)). The mean value of R=0.364, in the variation range 0.047 – 0.96 with a standard deviation of 0.215. For comparison with the threshold value of the operational process, experimental results on the crack development under harmonic loading in the form of fatigue failure kinetic diagram (FFKD) for the same steel grade were taken from [6]. Figure 4 (a) shows a diagram of 20GFL steel for the skewness coefficient R=0.5, as the closest value to the average value of the skewness coefficient of the operational loading process under consideration R=0.364 (figure 3(b)). The values of the skewness coefficient were calculated to determine the crack development rate variation pattern at the low level.

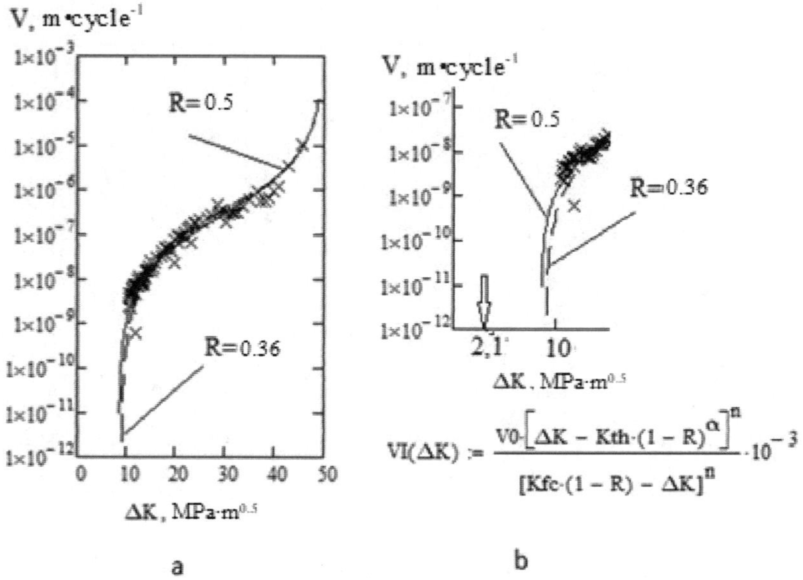

Figure 4. (a) — Kinetic diagram of fatigue fracture of 20GFL steel with a skewness coefficient of R=0.5 [6] and 0.36 with the extrapolation threshold level $\Delta Kth=8.95$ MPa·m$^{0.5}$ at R=0.36; (b) — Area thresholds of 20GFL steel (figure 3) defined with the parameters [6] (presented in the top part of the dependence graph) and operational process SIF value Kmax=2.807 and R=0.36.

The calculation was performed using the parameters defined in [6] for $R \leq 0.5$. This is Kth=12.34 MPa·m$^{0.5}$at R=0 (zero cycle) and $\alpha=0.846-0.37$ R. On the FFKD graph (figure 4 (a)), the calculated dependence at R=0.364 is shown as a dotted line and corresponds to the threshold value of the SIF amplitude $\Delta K=8.949$ MPa·m$^{0.5}$. For the operational threshold process, the same calculation of the SIF threshold amplitude, carried out on the average level of the maximum values of 2.807 MPa·m$^{0.5}$and the average value of R=0.364 determines the value of 2.13 MPa·m$^{0.5}$. Figure 4 (b) shows the enlarged lower part of figure 4 (a). It can be seen that threshold value of the SIF amplitude determined by extrapolation of the FFKD is significantly greater than those found from the average values of the operational threshold process (this value is shown by the arrow). Even for the maximum value, process values of 5.458 MPa·m$^{0.5}$, the threshold amplitude will be 3.96 MPa·m$^{0.5}$, which is almost half the threshold level defined by the FFKD (8.949). It shows that this type of operational process, such as [7], causes much more damage compared to simple harmonic loading.

Conclusions

The above analysis shows that the threshold level of development of fatigue cracks in low-alloy steel of 20GFL grade (specimens from the cast bolster of a freight car), obtained by extrapolating the fatigue failure diagram, is significantly higher than determined by the forecasting of the threshold level of the operational loading process. Thus, as per average value of maxima of the process 2.807 MPa·m$^{0.5}$and the average skewness factor R=0.364, the threshold SIF amplitude is equal to 2.13 MPa·m$^{0.5}$, and as per FFKD obtained under harmonic loading, this parameter is equal to 8.95 MPa·m$^{0.5}$. In other words, the use of operational process model [7] results in greater damage compared to harmonic loading. This should be taken into account in the calculations of such critical structural elements as the bolster and the side frame of a freight car, made of this steel grade.

References

[1] Makhutov N A, Moskvitin G V and Lebedinsky S G 2018 *Problems of strength, technogenic safety and structural materials science* ed N A Makhutov, Yu G Matvienko and A N Romanov (Moscow: Lenand)

[2] Romanov A N 2013 Propagation of fatigue cracks and a single curve of cyclic crack resistance of structural materials *Mechanical engineering and machine reliability studies* vol **5** pp 47-57

[3] Petit J, Henaff G and Sarrazin-Baudoux C 2000 Mechanisms and modeling of near threshold fatigue crack propagation *Fatigue Crack Growth Thresholds, Endurance Limits and Design* vol **1372** ed J C Newman and R S Piascik (West Conshohocken PA: ASTM)

[4] Sunder R 2015 Characterization of threshold stress intensity as a function of near-tip residual stress: theory, experiment, and applications, materials performance and characterization *ASTM Int.* vol **4(2)** pp 105-30

[5] Zheng X 1987 A simple Formula for fatigue crack propagation and a new method for the determination of ΔKth *Engineering fracture mechanics* vol **27(4)** pp 465-75

[6] Lebedinsky S G and Zmeeva V N 2000 Regularities of fatigue cracks development in cast steels of railway structures *Mechanical engineering and machine reliability studies* vol **3** pp 98-103

[7] Lebedinsky S G, Moskvitin G V, Pugachev M S and Polyakov A N 2020 Regularities of fatigue cracks development in steel at a low level of operational loading *Mechanical engineering and machine reliability studies* vol **2** pp 73-79

Acknowledgments

The authors would like to thank colleagues for important advices and assistance in preparing this article.

IOP Conf. Series: Materials Science and Engineering 1023 (2021) 012016 doi:10.1088/1757-899X/1023/1/012016

Threshold level of cast railway steel loading in the process of operation

S G Lebedinsky, G V Moskvitin and M S Pugachev

Mechanical Engineering Research Institute of the Russian Academy of Sciences, Moscow, 101990, Russia

GVMoskvitin@yandex.ru

Abstract. The authors studied the pattern of fatigue cracks distribution in railway J02002-type grade steel at threshold level of operational loading. The study was based on simulation of operational load combining the conditions of straight and curved part of the road. The load level reduction was based on the pattern of the change of the tested sample hardness with the development of the crack. The authors found the relation between the load level reduction, the crack growth rate and the number of load units. They determined the coefficient of similarity of operation loading form, at which there is no fatigue crack growth in the steel under consideration.

1. Introduction

There exists the problem of assessment of railway structures resource. In particular, it concerns the bogies, because in the process of manufacture thereof the defects of casting inevitably occur to become the stress concentrator that provokes the development of fatigue cracks. The resource of the bogies greatly depends on the duration of fatigue cracks growth to the critical values. The process of accumulation the damages at pre-threshold cyclical load depends on many factors. The main of them are: the cyclical hysteresis of stresses and deformations at the tip of the crack and its correlation with the residual stresses [1-3], as well as the impact of environment.

Experimental studies of threshold level assessment at operational load are too expensive and call for significant time expenditures, however, the authors suggest the method that provides the prediction of threshold level at the straight road section based on reduction of the sample hardness with the development of fatigue crack in it [4].

The task of this work is to develop the method that provides the assessment of threshold level of operational load by simulating a more complex loading process, in particular, the combination of conditions of straight and curved road section.

2. Test methods

The tests were performed at a compact C(T)-type 125x120x10 mm sample of J02002 steel. The loading pattern corresponded to ASTM E399. The sample was cut out of cast truck bolster of a railway car and had an initial fatigue crack Lo=42.0 mm, from the load application line. Crack opening indicator was installed on attachable prisms attached to the butt end of the sample. In the process of the test, the authors were constantly measuring the load applied to the sample and the crack edges displacement with registration frequency of 5000 values per minute for each channel.

The test method was as follows:

- operational loading, that is, an analogue of a truck bolster loading process recorded during operation, was applied to the sample. It was performed in mild loading mode. Alongside with it, the sample response and the change of crack advancement speed V were recorded from the crack opening indicator. The length was measured using the marks applied on the polished side surface of the sample;
- after the crack growing ΔL by 0.5 mm and recording its initial speed, the sample loading was transferred to the severe stress mode. A signal previously recorded from the opening indicator, when the loading was performed in a mild mode, was used to control the crack edges displacement. In the process of severe loading, the sample response was registered from the load change indicator. At the first period of severe mode the loading signal completely corresponded to the mild mode operation signal, however, with the crack growth, the sample compliance ΔP increases, and the load level decreases. It was necessary to determine the change of load cycle parameters correlation. By doing so, the following dependencies were determined: the crack length change with the number of repeated blocks and the stress level reduction according to the crack length.

The rate of loading at servohydraulic test machine was determined by the types of operational loading load cycles, as well as the accuracy of reproduction thereof. In this experiment, it constituted 3 cycles per second (figure 1).

In the mild loading mode, the initial crack speed was 22.8×10^{-7} m per block during reproducing the operational loading block corresponding to the rolling stock speed of 80-90 $km\cdot h^{-1}$ along the straight and curved track sections.

Using the current loading readings in figure 2 (a), the displacement of point 1 was tracked in the graphs as a value corresponding to the maximum load in each block. The displacement of point 1, at crack growth figure 2 (b), was determined according to the crack growth value ΔL and the number of worked out blocks. The ΔL value was measured using a counting microscope.

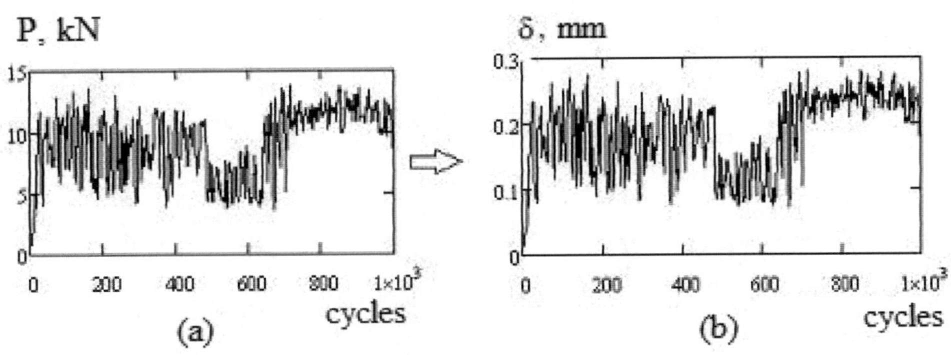

Figure 1. Type of operational loading block. (a) - in mild mode (loading control), (b) - under severe loading (crack opening sensor control).

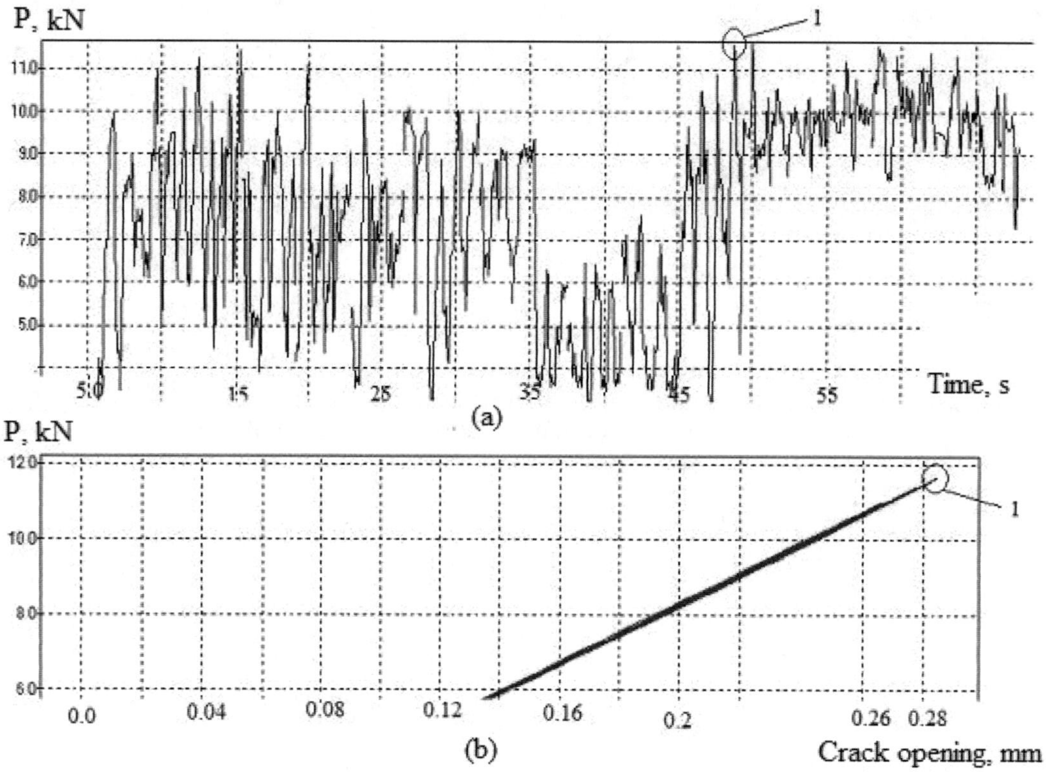

Figure 2. Operational loading blocks during the movement of the car along straight and curved track sections in coordinates. (a) - load – time, (b) - load - crack edges displacement at the opening sensor installation points.

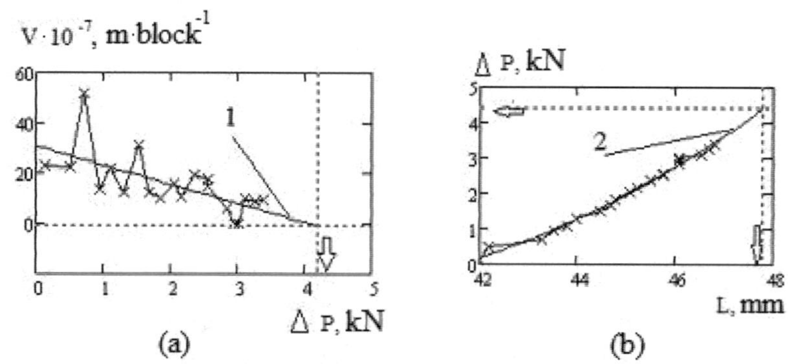

Figure 3. Experimental functions. (a) - dependence of the crack growth rate on the decrease of the maximum loading in the block, (b) - dependence of the decrease of the maximum loading in the block on the crack length, 1 and 2- approximating experimental points of the function.

3. Research results

Based on the test results, we obtained the experimental dependences of the change in the specimen compliance in the form of functions ΔP (ΔL), and the speed change with the decrease in the loading level V (ΔP) and performed the calculation assessment of the interval and regularity of the stress intensity factor (SIF) change taking into account all the experiment parameters (figure 3).

During the tests, the crack increased from 42 to 46.06 mm, while the maximum loading level decreased from 13.74 to 10 kN.

SIF was determined according to ASTM E399 for a compact sample.

The calculated SIF dependences in the test section of the sample only on the crack length change are shown in figure 4 (a). Determination of this dependence provides the assessment of the possible experiment duration and makes it possible to choose the initial conditions. Figure 4 (b) shows the distribution of the values of the cycles asymmetry coefficient, determined from the adjacent minimum and maximum values of the loading process. With the decrease in the average process level of the in the experiment, the sequence of the cycle asymmetry coefficients does not change.

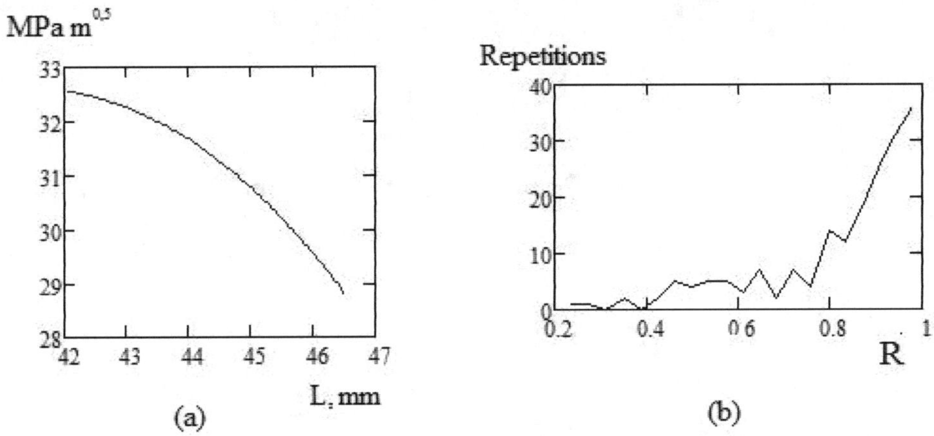

Figure 4. The calculated SIF dependences. (a) - on the crack length in the experimental section of the sample at a constant load, (b) - distribution of asymmetry coefficients in the cyclic loading block.

The analysis of the loading process during its level decrease was carried out according to the ratio of the limiting values of the loading blocks amplitudes for mild and severe modes. It has been determined that with a decrease in the loading level with the development of a crack, the ratio of the extreme values of the loading cycles remains constant. That is, the ratio of the consecutively standing in the cycle of the minimum value to the maximum value remains constant. This serves as a measure of the observance of the similarity when the load level decreases. The loading amplitude changes accordingly.

The analysis of the loading process during its reduction was performed based on the correlation of the threshold values of load units amplitude for mild and severe mode. It is determined that during the loading level reduction with the crack development, the correlation of loading cycles extreme values remains constant. That is, the correlation of the minimal value to the maximum value, subsequently located in the cycle, remain constant. It serves as the measure of compliance with similarity during the load reduction level. At this, the loading amplitude correspondingly changes.

Using line 2 figure 3 (b), it is possible to obtain the threshold loading process assessment, expressed in the sequence of SIF values, corresponding to extreme values of cycling loading and the predicted crack value. Therefore, the threshold crack length, obtained this way under the experiment conditions and the loading level, corresponding to it, with the maximal loading in the unit equal to 9.61 kN, at similarity coefficient equal to γ =0.70, make it possible to obtain the threshold unit loading assessment in the form of the corresponding sequence of cyclically changing SIF at which the crack terminates.

The possibility of obtaining the threshold process by multiplying each initial process extremum by similarity coefficient is shown in figure 5.

The initial loading process in SIF values sequence is shown in figure 5 (a). The predicted process of SIF changing corresponding to crack termination condition is shown in figure 5 (b). The coefficient of similarity of initial and predicted threshold process is 0.7, that is, the predicted process is obtained by multiplying the initial processes extremums by similarity coefficient.

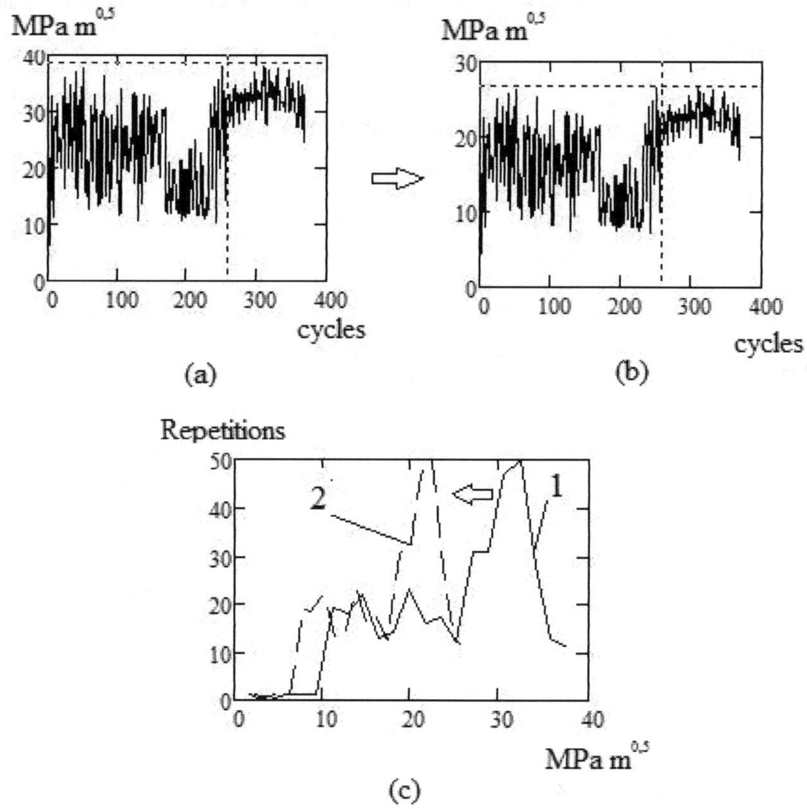

Figure 5. Obtaining SIF change threshold process. (a) – source process, (b) – predicted process, (c) – histograms of initial (1) and threshold (2) processes broken down by amplitudes spectra.

4. Conclusions

The authors have revealed the pattern of fatigue cracks development in J02002 steel, of which the freight car truck bolster is manufactured, based on experimental study of a compact sample. The authors used the records on operational loading specific for combination of straight and curved railroad sections.

The authors have developed a method to determine the threshold level of operational loading at such loading process type.

It has been established that, within the experiment conducted, the loading process implementation in the units remains similar regardless the loading mode, crack growth and loading level reduction.

Using the similarity coefficient, it was possible to determine the change of SIF similar to operational one at which the crack development terminated.

The analysis of the sample surface treatment with a heated air flow and a mechanical mixture of metal and corundum particles indicates the effect of copper particles deformation on the predominant diffusion of copper into zinc and the increase in the diffusion coefficient.

References

[1] Petit J, Henaff G and Sarrazin-Baudoux C 2000 *ASTM Special Technical Publication* **1372**
[2] Sunder R 2005 *Fatigue & Fracture Engineering Materials & Structures* **28** pp 289-300
[3] Sunder R 2015 *J. of ASTM Int.* **4** pp 105-30
[4] Lebedinsky S G, Moskvitin G V, Pugachev M S and Polyakov A N 2020 *J. of Machinery Manufacture and Reliability* **49** pp 144-49

IOP Conf. Series: Materials Science and Engineering 1023 (2021) 012017 doi:10.1088/1757-899X/1023/1/012017

The development of models of risk assessment complex transport systems

N A Makhutov, A N Panov and O N Yudina

Mechanical Engineering Research Institute of the Russian Academy of Sciences, 4 Maly Kharitonyevsky Pereulok, Moscow, 101990, Russia

a.panov@tut.by

Abstract. The article deals with the actual problem of risk optimization for complex transport systems. Optimization models for risk aspects are proposed and developed: "probability-damage"; "quality-price"; "variation-damage"; acceptability of the measurement system; degradation monitoring.

The urgent task of our time is to increase effectiveness and efficiency of creation and functioning in τ time of complex social engineering and natural engineering systems. This fully applies to complex defense and civil transport wheeled vehicles that are used in extreme conditions (figure 1). This problem can be resolved through analysis, management optimization and risk optimization, the effects of uncertainty [1, 2]. A more detailed risk analysis in the correlation between $P(\tau)$ probability of a hazardous event and $U(\tau)$ damage because of this event, between variation and damage with optimization of the product/process according to the price-quality criteria [3] made it possible to develop an appropriate fundamental model (figure 2).

Figure 1. Mobile machines and elements that determine their safety.

$P(\tau)$ parameters are determined with hazardous process and operational influences, and $U(\tau)$ values depend on the liability and cost of the machines, as well as the degree of damage caused by impacts.

It should be noted that the probability-damage risk analysis and risk management start from a point single-criterion assessment and develops step by step to implementation of the prioritization strategy in the form of hyperbolic dependence. For easy practical use, the nonlinear function is converted into a stepped point scale with the inverse approximation (figure 2). The proposed model makes it possible to harmonize the criteria for achieving compliance with the established requirements for complex engineering systems, including transport, based on the risk-based approach. This allows you to optimize costs when creating a type of engineering object, its replication and operation based on prioritization for the type of the engineering system, to set goals in the scope of compliance variation control during replication, depending on priorities, based on identification of the optimal quality-price ratio function (figure 2).

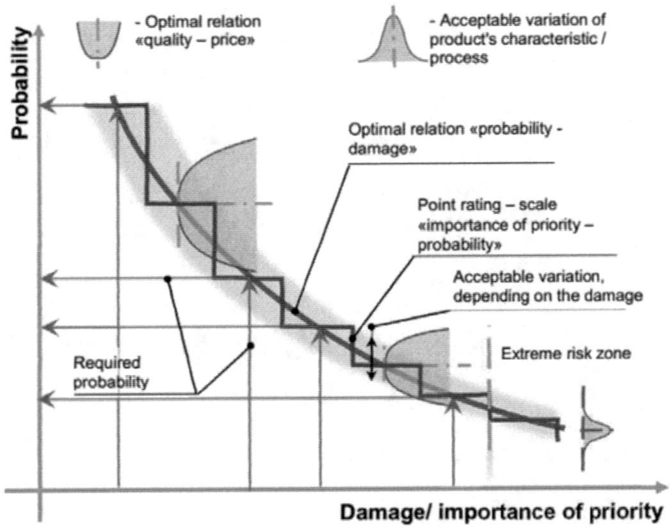

Figure 2. Fundamental five-component model of optimization of dependencies "probability-damage" – "quality-price" – "variation-damage".

Depending on the potential damage, the criteria are proposed for optimizing costs for adequate interrelated accuracy of risk assessment, depending on the priorities, fundamental models for the choice of measuring systems used both for creating a type of engineering object and for its replication at: selection of types of measuring systems; tolerance for the acceptability of variation of arrays of measuring systems for the selected type (figure 3).

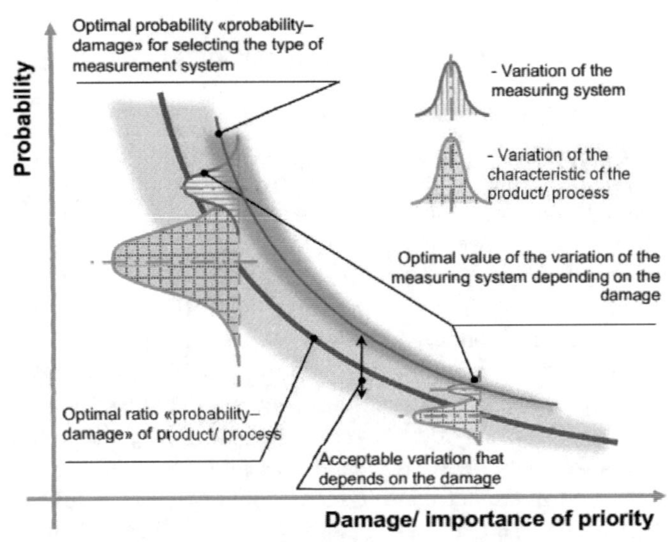

Figure 3. Fundamental model of optimization of the choice of the type of measuring system and its variations depending on the damage.

To reduce the cost of creating engineering objects by using means of the V process order, a risk optimization model has been developed by introducing and applying a subsystem of diagnostics/monitoring of degradation in the social-engineering system (figure 4).

This allows you to reduce unnecessary safety factors and the failure probability without detriment to the consumer. If there is evidence that the degradation process of an engineering system is approaching a critical value, the excessive risk, the engineering system prevents occurrence of this event and allows you to move to the acceptable risk zone (figure 4).

Thus, it is possible to reduce the material consumption of an engineering object and operating costs with acceptable failure risks when it is functioning [3, 4]. It should be noted that the risks of measuring systems for identification of degradation processes, depending on the risks of the engineering systems they serve, are also taken into account, both when choosing acceptability of the type, and when assessing acceptability of variation of the measuring systems (figure 4).

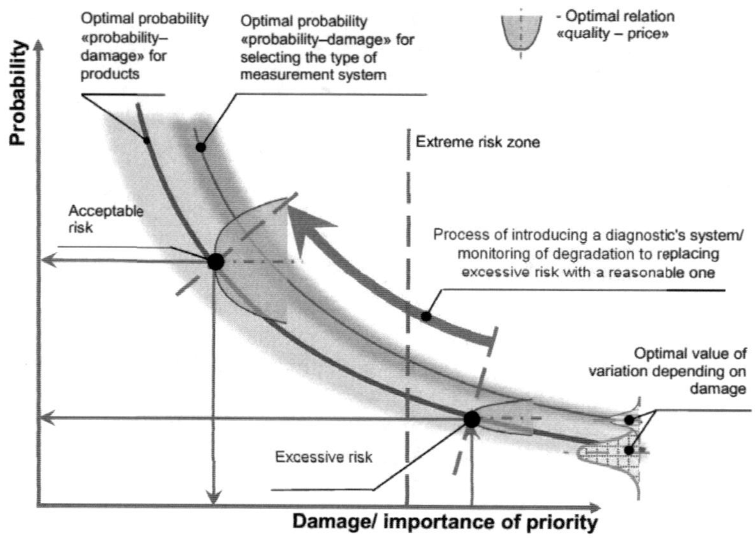

Figure 4. Model for risk optimization by introducing a system for diagnostics/ monitoring of degradation.

The proposed models were gradually applied in the development of technical regulatory legal acts of state and interstate standards to scale the economic effect based on widespread using of risk-based thinking [1-4] when creating equipment and technologies by optimizing costs to reduce the cost of products and processes to competitive values while meeting consumer requirements and legal restrictions.

For legislative, systematic application in organizations that design and produce complex engineering systems, the proposed models were used for development of standards for design and production management systems for the following: mobile machines of various types (road, quarry, automotive and other equipment) [5]; special dual-purpose equipment [6].

Numerous methods of risk analysis for both products and processes have different degrees of reliability and resource requirements, including databases and knowledge bases [1-4], while FMEA method is the most widely used in the automotive industry. To prevent its inappropriate practical use, it is proposed [7] to use risk management plans, cause-and-effect trees (deduction/induction), etc. as input data, which allows, based on the model of effective knowledge management developed in [3], to carry out decomposition and managing the likelihood of damage from a configurator (a top-level system, failure of which determines the damage, e.g., a hazardous chemical transportation system) to the root cause of a nonconformity (e.g., chemical composition of the material of the part) based on the present level of knowledge.

At present, numerous methods have been developed and are used for assessing the variation of product replication processes, statistical process control, statistical acceptance (streaming, batch), as well as continuous control [8]. Their systematization and methodological support are required for their adequate application in the risk-oriented approach according to the variation-damage criterion (see figure 2). To select an acceptable variation that is based on the quality-price methodology (see figure 2), the methodological support for selection of appropriate statistical methods was proposed for the first time, both depending on potential damage and taking into account the stages of verification and validation processes in preparation of manufacturing of products and at daily adjustment of the manufacturing process [8].

It was proposed [9] to improve the methods of analysing the variation of measuring systems both in methodological terms (taking into account numerous factors, e.g., personnel, environment, type of measuring system), and in combination with the risk-oriented approach of their application, depending on the potential damage [9].

Methodology of risk management within implementation of projects in the supply chain is proposed [10] to ensure the probability of non-conformances that appear both at designing and development of a type of product and manufacturing processes, and at its replicating, through the whole supply chain (starting from creating of materials, parts, units, finished products, maintenance/recovery to disposal).

Methodological support for risk analysis, management and optimization has been created, including one based on developed risk-oriented models, for wide practical application when creating engineering systems for adequate risk analyzes [7], statistical methods [8], measurement processes [9], verification and validation in the supply chain from primary suppliers of materials to suppliers of maintenance processes in operation and disposal of engineering products [10].

Interstate standards have been developed to implement the optimal requirements in designing and manufacturing of components of the transport systems that primarily ensure their active and passive safety using the proposed models, which, among other things, are used in development of the EAEU technical regulations (TR TS 018 EAEU) for steering, brakes, suspension and supporting structures of vehicles [11-14].

References

[1] Abrosimov N V *et al* 2015 *Safety of Russia. Legal, Social, Economic, Scientific and Technological Aspects. Scientific Basis of Manmade Safety* ed N A Makutov (Moscow: Znanie publ) p 936

[2] Abrosimov N V *et al* 2018 *Safety of Russia. Legal, Social, Economic, Scientific and Technological Aspects. Manmade, Technological and Aanthropogenic Safety* ed N A Makutov (Moscow: Znanie publ) p 1016

[3] Gorbatsevich M I, Panov A N and Minyukovich S M 2005 *Design of Vehicles: Loading, Damage, Resource* ed A N Panov (Minsk: "Technoprint") p 264

[4] Alekseev V V, Gavrilenko S I, Panov A N and Stolyarov V E 2016 *Mathematical Models and Methods for Ensuring Acceptable Risks of Information-measuring and Control Systems of Transit Gas Pipelines* (St. Petersburg: Publishing house of SPbGETU "LETI") p 160

[5] STB 16949-2018 Quality management systems. Particular requirements for the application of STB ISO 9001-2015 for the organizations involved in the supply chains of Automobile and Tractor, agricultural, material handling, career and special machine building

[6] STB V 15.004-2009 System for development and production of defense products. Military equipment. Quality management system. Requirements. Enter. 01.01.2011 (Minsk: State Standard)

[7] STB 1506-2015 Management systems. Risk management. Failure mode and effects analysis method

[8] STB 1505-2015 Management systems. Processes management. Methods of statistical process control
[9] STB 2450-2016 Management systems. Measurement management. Management system analysis
[10] STB 2484-2016 Management systems. Management. Verification and validation in the supply chain
[11] GOST 30729-2001 Joint ball-pines, ball-joints of rod ends of vehicles suspension and steering. General technical requirements, acceptance rules and test methods
[12] GOST30731-2001 Cylinders, tubes and hoses of hydraulic drive of brakes and clutch of vehicles. General technical requirements, acceptance rules and methods of control
[13] GOST 31341-2007 Brake pads, discs, and drums for vehicles. General technical requirements, acceptance rules and test methods
[14] GOST 30551-98 Trunks for installation on the roof of the body (cab) of the vehicle and trailer. Technical requirements, acceptance rules and test methods

Monitoring of the NPP equipment stressed state according to indirect measurements data

S V Maslov[1] **and A I Kukshinov**[2]

[1] Mechanical Engineering Research Institute of the Russian Academy of Sciences, 4 Maly Kharitonyevskiy Pereulok, Moscow, 101990, Russia
[2] Nizhny Novgorod Engineering Company Atomenergoproekt (Branch), 2/1 Dmitrovskoe Shosse, Moscow, 127434, Russia

maslovsv@inbox.ru

Abstract. The method for monitoring the stress state of the NPP equipment internal surfaces is considered, based on the results of strain gauge measurements obtained for the outer surface of the structure. The temperatures of the internal surfaces are found from thermoelasticity inverse problem solutions, the stress state is determined by the solution of direct problems.

1. Introduction

The provision of safe operation of pressurized water reactor (PWR) nuclear power plants (NPP) is associated with the use of diagnostic systems for indirect control of loading of the elements subject to high thermal exposure on the part of heat transfer media with changeable temperature. The use of such diagnostic systems is necessitated by the major deviation of the real conditions of equipment thermomechanical loading from the rated models accepted at the stage of design strength calculation. The objective of creation the algorithms and systems of stress-strain state (SSS) control in switchgear equipment is to arrange the processing of experimental and experimental – computational data about the processes of real loading of the structure elements and units during its operation for control and accounting of the damages accumulated in metal. The algorithms of operative information collection and processing that provide data about stressed state and accumulated damages in the points inaccessible for direct measurements, may rely on the results of indirect measurements and mathematical modelling of thermomechanical loading processes in the structure [1]. The methods of inverse heat conduction problem solution, that, however, in many cases do not provide stable solutions, are typically used for modelling of thermal exposure of heat transfer media from the internal surfaces of the structure. In this paper it is suggested to use the iteration method of inverse heat conduction problem solution to build the model of thermal exposure on the structure from its internal surface [2]. In this case, the results of strain gauge measurements obtained for external surface, are used as the initial data.

The following assumptions are used to form the algorithm providing the operative building of iteration process and the search for solution corresponding (within the measurement errors) to the experimental data obtained.

1. Heat transfer coefficient from the heat transfer media to the structure wall is constant all over the considered section of the internal surface and does not change over time.

2. Due to the efficient heat insulation, there is no heat transfer on the external surface of the structure.

As we can see from the experience of the tests conducted [3, 4], the above-specified assumptions correspond to the real conditions of NPP equipment operation [5].

2. Research method

As we can see from the experience of the tests conducted [3, 4], the above-specified assumptions correspond to the real conditions of NPP equipment operation [5].

To implement the suggested algorithm, it is necessary to exclude the components of stresses in the dangerous section, caused by the bend of the attached parts of the structure (pipelines) during irregular change of their temperature, from the initial data of thermoelasticity inverse problem. Unfortunately, the direct experimental determination of these stresses is not possible. Besides, the computational determination of the above-specified stress components does not provide the results of acceptable accuracy due to uncertainty of heat exchange conditions and the complex geometry of pipelines system that contains curvelinear sections with variable temperature. Therefore, it is suggested to use the specialized diagram of additional sensors installation at the direct section of the attached pipeline, at the distance from the stress concentration zone. The example of such specialized diagram of strain gauge and thermometrical control sensors is shown in figure 1.

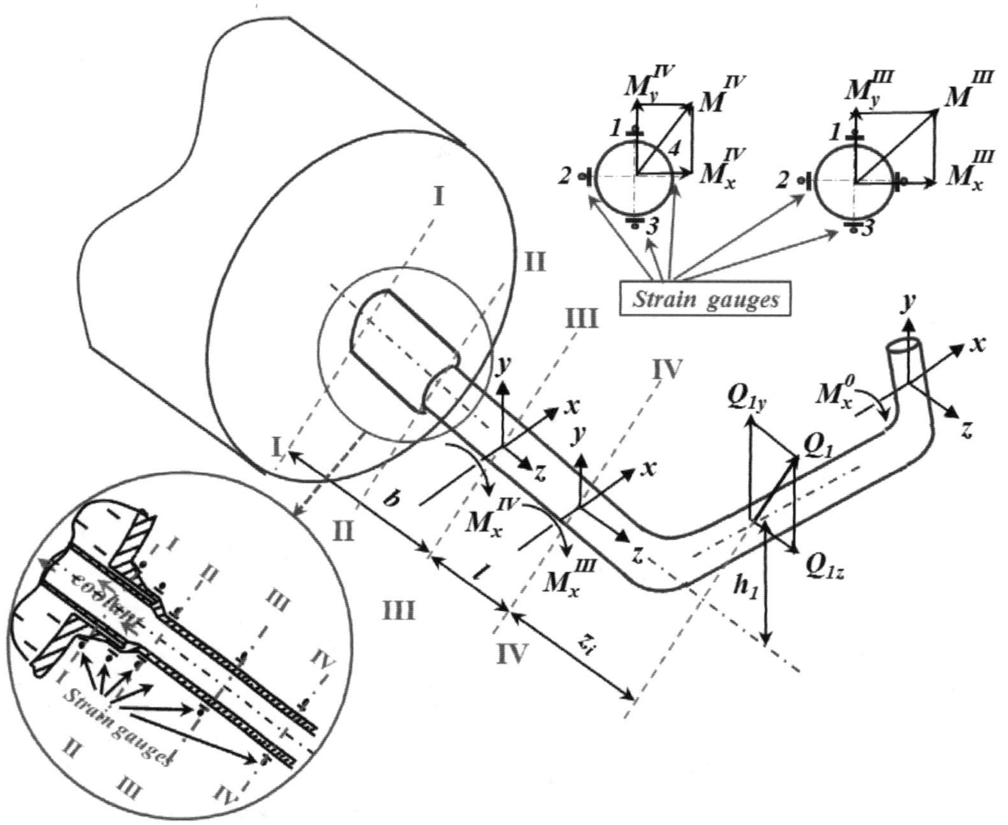

Figure 1. Installation diagram of the monitoring system sensors designed to determine the stress state of the VVER steam generator coolant injection unit.

Strain gauges installed in pipeline sections III - III and IV - IV, can be used to exclude the effect of supplementary bending moments on the stresses measured in the dangerous sections I - I and II – II, as is shown in figure 1. At least 4 strain gauge sockets are installed in each of these sections which provide the determination of the value and direction of the bending moment acting in the bending moment section. The differences of the axial stress component, determined by the indications of the sensors, are located at the opposite sides of the pipe section. Besides, the figure 1 shows the diagram explaining the origin of the additional bending moments and the order of determination thereof for the dangerous sections I – I and II – II. From the diagram provided in the figure it is seen that the bending moments occurring in the sections under consideration, depend on the points of application and the values of external loadings – the transversal forces Q_i and the external moments M_x^O. For example, the bending moment in the section III - III is determined from the following equation:

$$M_x^{III} = \sum_{i=1}^{n} Q_{iz} h_i - \sum_{i=1}^{n} Q_{iy} (z_i + l) + M_x^O \quad (1)$$

The differential bending moments in the controlled sections III – III and IV – IV depend only on the axial distance between them:

$$M_x^{IV} - M_x^{III} = \sum_{i=1}^{n} Q_{iy} l \quad (2)$$

In the equation (2), the moment values are known as they are determined from the tensoresistors readings. Therefore, the unknown moment in the dangerous section II – II can be determined from the formula:

$$M_x^{II} = M_x^{IV} + \frac{l+b}{l}(M_x^{IV} - M_x^{III}) \quad (3)$$

The formulas for determining the bending moments M_y are similar to equations (1) - (3).

The obtained functions of the change of stresses at the external surface of straight pipeline section are used to build the function of temperature variation of heat transfer media producing heat exposure on the internal surface of the structure. For this purpose, the continuous function of thermal variation in the heat transfer media is replaced with the step function with time increment $\Delta\tau$, shown in figure 2. To determine the relation of the heat transfer media temperature variation with the stresses, the function ψ, determined by calculation, is introduced; it is the function of the change of stresses on the external surface during stepwise temperature change in the heat transfer media by 1 degree. The function ψ is determined by numerical calculation in the software complex ANSYS for real geometry of straight pipeline part at established single temperature pulse and the calculation. To build the function with the temperature variation, the differentiable function $\psi(\tau)$ is substituted with discrete function, as is shown in figure 3.

Building the function ψ it is necessary to consider that its value should exceed the possible error of strain gauge stress measurement [6]; otherwise, the solution obtained will be unstable. Therefore, at the sections where this condition is not fulfilled, the ψ value is taken equal to zero, as is shown in the figure 3. Besides, it is important to understand that the values of time intervals $\Delta\tau$ during the temperature variation function T schematization (as shown in figure 2) and the functions of stress response to temperature rise (figure 3) should be equal.

The resolving equation for determination of heat transfer media change variation has the following form:

$$\sigma_i^* = \psi_i \Delta T_1 + \psi_{i-1} \Delta T_2 + \ldots + \psi_2 \Delta T_{i-1} + \psi_1 \Delta T_i \quad (4)$$

The following indications are used in the formula (4): σ_i^* - the measured stress values exclusive of the components caused by the bending moments and the normal forces acting in the sections under consideration; ΔT_i and ψ_i are correspondingly the increment of the heat transfer media temperature and

IOP Conf. Series: Materials Science and Engineering 1023 (2021) 012018 doi:10.1088/1757-899X/1023/1/012018

the value of ψ function at the sections *1, 2, ... i*, where "*i*" is the number of the measured temperature stresses.

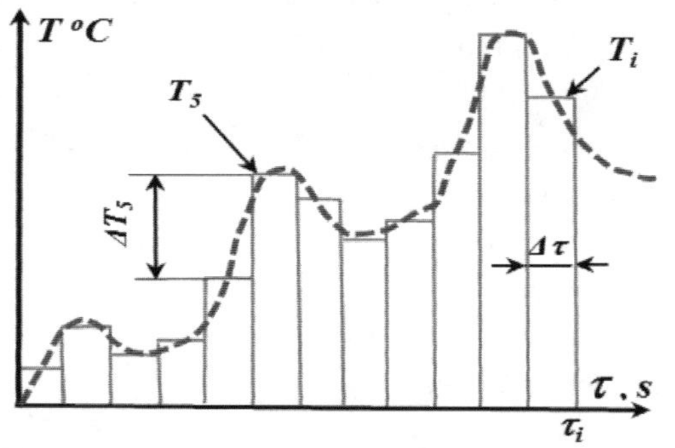

Figure 2. Schematization of coolant temporal variation function.

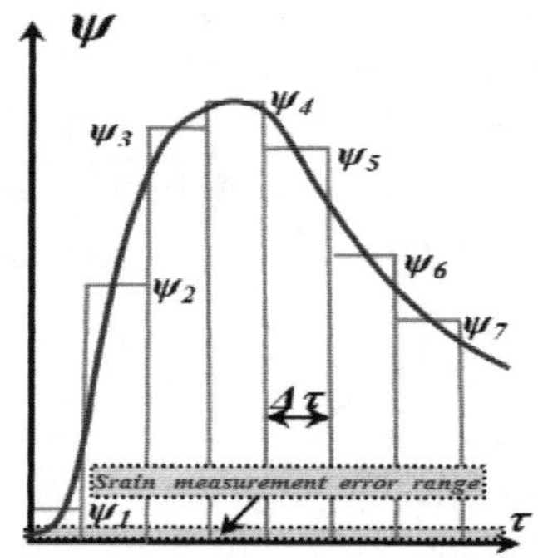

Figure 3. Schematization of the function ψ – temporal stress response σ to a single stage of temperature rise.

The algorithm based on consequential solution of equations (4), developed in this work, consists of the following steps.

1. Determine the range of the possible values of heat transfer coefficient from the heat transfer media to internal walls of the structure from supplementary computational and available experimental data. As we can see from the numerical modelling performed, the error of heat transfer coefficient setting does not add major errors to SSS state monitoring results. It is due to the fact that the intensity of thermal exposure of the internal surface is determined by the product of heat transfer coefficient and the temperature difference between the surface and the heat transfer media. Therefore, the mistake in determination of heat transfer coefficient is essentially compensated by the corresponding temperature variation determined by the algorithm suggested.
2. Based on the measured values of stresses σ_1^* determine ΔT_1 value for the first step by time from the formula (4), containing only one component in the right part.
3. Determine the values ΔT_2, ΔT_3 from the type (4) equations, containing, correspondingly, 2 and 3 components in the right part.
4. Consider type (4) equation for i=4, but ψ_1 value is taken equal to zero.
5. Consider the further iterations, where the function ψ_i values are taken into consideration only for three previous iteration steps.
6. The staged temperature variation function is interpolated by the smooth function obtained by spline-interpolation method.

The temperature variation function, determined as above, is used for setting the limiting conditions of 3rd type all over the internal surface of the pipeline and the duct attached to it (figure 1). In case that the internal surface of the joint of the pipeline connection to the main body has an annular space as shown in figure 1, in this space the temperature is taken invariable or slowly changing according to the linear law. This assumption results in slightly overevaluated calculated temperature stress values, which increases the safety margin of the structure under consideration. The choice of a particular function of temperature variation in the annular space is based on the best approximation of the values of computational temperatures and stresses to the readings of strain gauge and thermometrical sensors in the section I – I.

To determine the temperature fields and the stressed state in the dangerous points, it is advisable to use the algorithm of direct heat conduction and thermoelasticity problems solution by finite elements method, for example, in ANSYS media. To set the initial data in direct tasks solution, it is necessary to use the obtained functions of heat transfer temperature variation and heat transfer coefficients, taken in the process of modeling thereof.

The calculation conducted is verified by comparison of the obtained computational data with the readings of measurement transducers installed in immediate proximity to the dangerous points of the structure shown in figure 1.

Figure 4 shows the results of computational-experimental determination of the steam generator feed water duct stressed state obtained by implementation of the algorithm suggested in this work. The temperatures (curve 2) and the circumferential stress (curve 3) on the internal surface of the joint of heat transfer media entry to the steam generator body are determined by the measured temperatures (curve 1) and circumferential stresses occurring on the external surface of the structure (curve 4). The correctness of the results obtained is confirmed by the proximity of these calculations to the results of the experiment conducted at the period of real operation of PWR-1000. The divergence of the calculation and experiment results does not exceed the measurement error that constitutes about 7%.

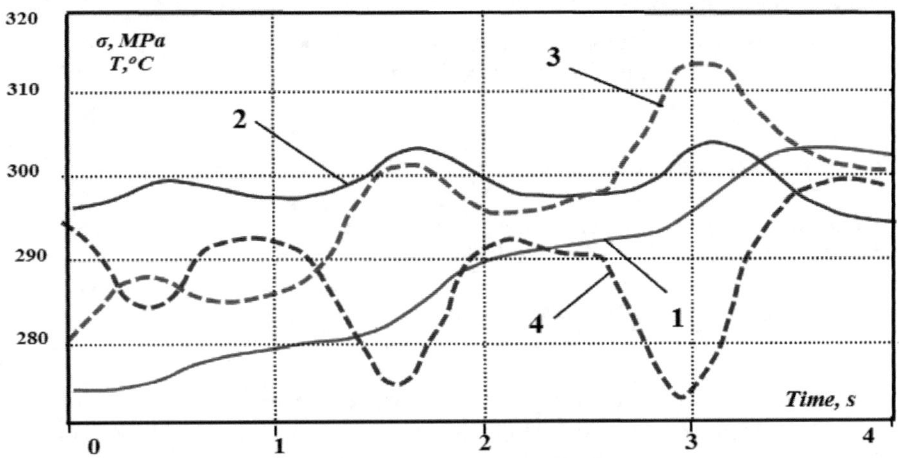

Figure 4. Measured and computational functions of temperature and stress variation in the area of PGV-1000M inlet branch pipe.

3. Conclusions

As the analysis result an algorithm is proposed for determining the stress state on the inner surfaces of NPP pipelines, based on solving inverse problems of thermoelasticity. It is shown that the measurement results obtained for the cylindrical sections of the pipeline remote from the geometric inhomogeneity sections and stress concentration zones may be used as the initial data. The proposed method is based on the accepted assumptions about the heat transfer coefficients from the coolant to the internal structural surfaces constancy. The developed algorithm is used to assess the stresses and temperatures in the feed water inlet zone of the VVER-1000 steam generator.

References

[1] Makhutov N A 2014 *J. of Machinery Manufacture and Reliability* **3** pp 217-32
[2] Kozlov V A, Maz'ya V G and Fomin A V 2009 *Computational Mathematics and Mathematical Phys.* **3** pp 525-31
[3] Razumovskii I A, Chernyatin A S and Fomin A V 2014 *Inorganic Materials* **15** pp 1528-15
[4] Mikhalev Y K, Fomin A V and Maslov S V 2008 *J. of Machinery Manufacture and Reliability* **5** pp 517-21
[5] Maslov S V 2019 *J. of Machinery Manufacture and Reliability* **2** pp 141-48
[6] Klymov M V, Maslov S V and Poguliaiko A N 2020 *IOP Conf. Series: Materials Science and Engineering. International Conference of Young Scientists and Students "Topical Problems of Mechanical Engineering"* ToPME 2019 C 012021

IOP Conf. Series: Materials Science and Engineering 1023 (2021) 012019 doi:10.1088/1757-899X/1023/1/012019

Development of models for risk analysis and optimization of risk reduction costs

A N Panov, D O Reznirov and O N Yudina

Mechanical Engineering Research Institute of the Russian Academy of Sciences, 4 Maly Kharitonyevsky Pereulok, Moscow, 101990, Russia

a.panov@tut.by

Abstract. The article deals with the actual problem of risk analysis and management in the technosphere. Proposed and developed: classification of risks (events, variations, bifurcations); risk analysis structures; risk management structure of variations and bifurcations; models for optimizing the cost of risk management.

Complex systems that include components belonging to technical, natural and anthropogenic environments are subjected to risks R induced by uncertainty. This is the most pressing problem of our time, in particular with the reduction of the resource base and nonlinear complication of the interaction between man and the manmade environments [1-3]. New national and international standards on risk management are being created [4-6]. Certain aspects in these standards are not sufficiently substantiated by the scientific and methodological basis [1, 2]. A more detailed study of the risks $R(\tau)$ allows introducing the classification of risk that takes into account the factor of time τ:

- risks of hazardous events (that are slightly dependent or conditionally independent of time);
- risks that change significantly over time, i.e. risks - process variations;
- risks with dramatic dynamic changes over time, i.e. risks with the effects of bifurcation and with the rapid reaching critical levels of risks $R_{\kappa}(\tau)$ (figure 1).

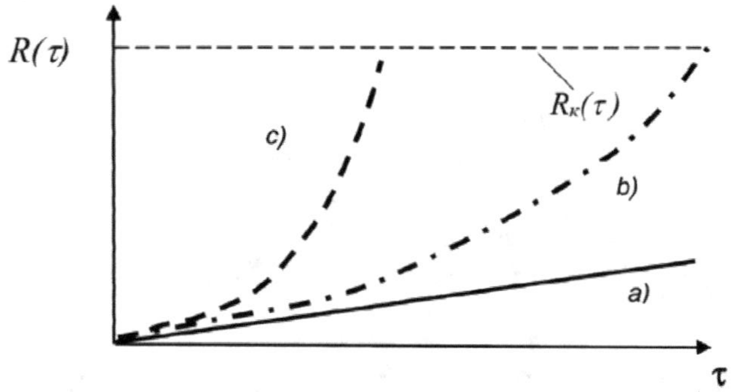

Figure 1. Classification of risks as a function of time a) risks of dangerous events, b) risks with significant changes over time, c) risks with large dynamic changes over time.

The following architecture of risk analysis is proposed to develop the existing models [1-3] (figure 2):

- identification of threats or an array of possible events (processes with non-robust variation; processes accompanied by bifurcations);
- determination of system vulnerabilities (taking into account the factor of time, figure 1);
- loss estimation.

Sociotechnical systems respond (figure 2) to threats, passing into the following states of vulnerability: stable, uncertain, unstable, overly unstable (response to a hypothetical extreme threat). Due to the inconsistency of the socio-technical system with threats the following losses can be inflicted: negligible, acceptable, excessive and hypothetical (figure 2).

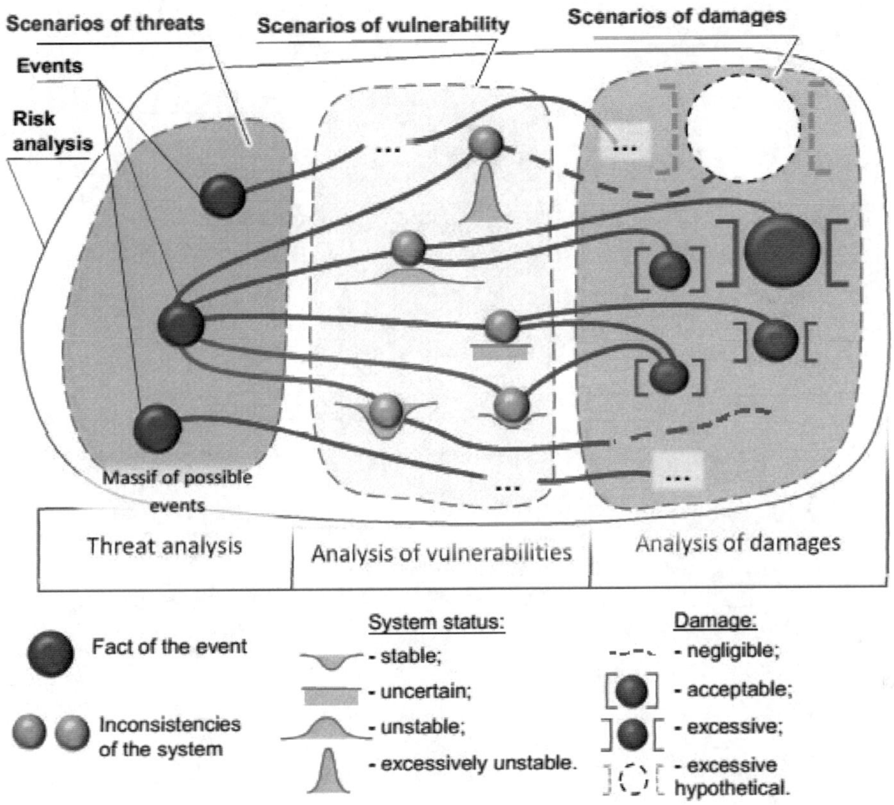

Figure 2. Risk Analysis Framework – Events.

Appropriate socio-technical solutions are implemented to prevent and counter risks [1-3]. Due to the negligible response time risks of bifurcations present a particular threat [1, 2]. A methodology for preventing and parrying bifurcations of processes is proposed. It can be illustrated by a graphical model (figure 3).

During the implementation of the process for which the requirements are established, on the basis of the risk analysis, the causes are determined and actions are taken to prevent, and in the event of unacceptable variations and bifurcations of parrying, as well as the intended actions are verified and validated.

If the arising bifurcations cannot be prevented by the provided preventive actions, in order to achieve the required output of the process with an acceptable nominal time value and its variation, an emergency action plan is implemented, after the implementation of which the acceptability of variation and prevention of bifurcations is achieved. Re-verification and validation allows the achievement of appropriate process output requirements despite bifurcation and satisfaction of process output customers and other interested parties (figure 3).

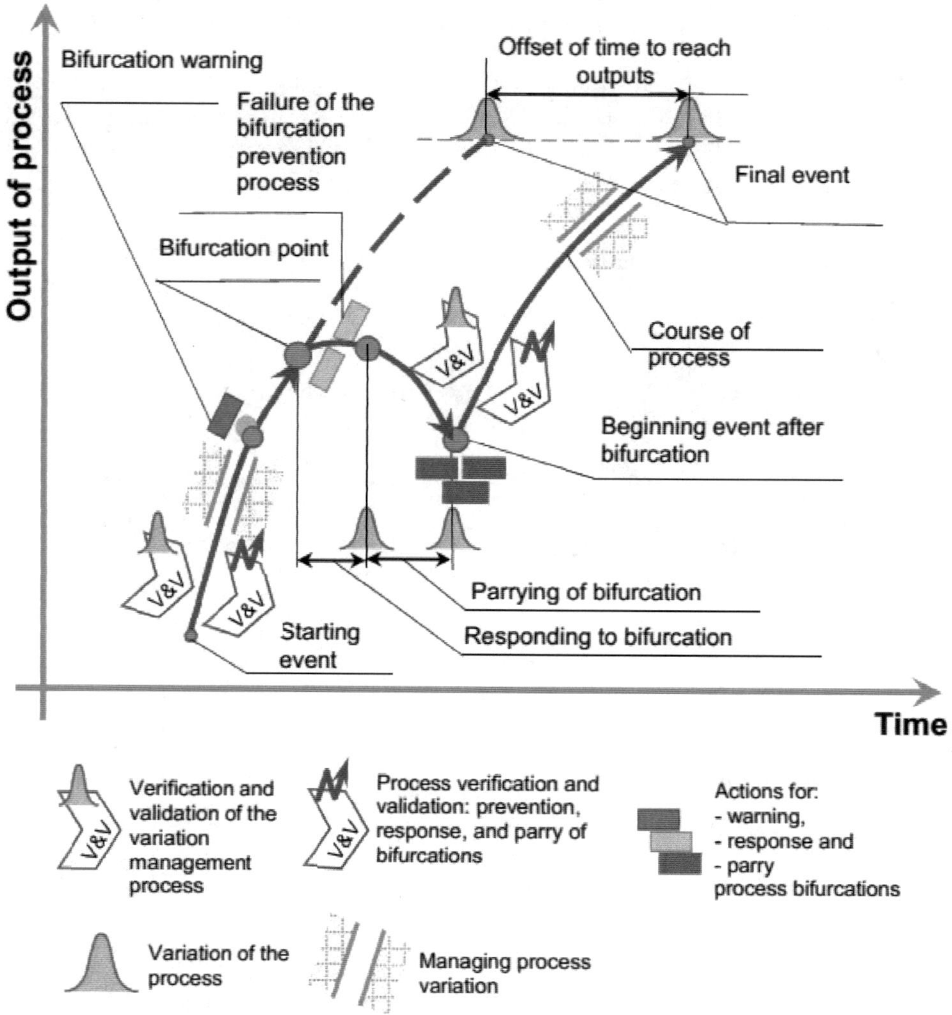

Figure 3. Process risk management structure by methods of prevention and parrying taking into account variation and bifurcation.

The optimization of costs is the most urgent problem in the implementation of risk-oriented thinking in practice [1, 2]. This is due to the fact that the costs of risk management have arisen as a new resource component.

To reduce the cost of products and processes, it is necessary to establish the optimal ratio between the types of costs and losses (figure 4). A model of cost optimization for achieving acceptable risks was proposed [3] based on the assumption of a close to hyperbolic relationship between each pair of components of identified groups of costs for achieving compliance in risk-oriented thinking. In this case, the minimum value of total costs is achieved when the three identified components are equal.

When implementing projects to create products, a methodological problem of adequate application of risk analysis techniques arises [3]. Technicians that perform the technical risk analysis within the traditional approach are not informed about strategic plans at the macro level. For example, they may not be aware of both the prospects of a particular project and plans for allocation of resources for implementation of measures to achieve the acceptability of risks [3]. To improve the risk analysis methodology, it is necessary, first of all, to introduce $R(\tau)$ risks for systems or components into the analysis, failure of which results to the maximum damage [7]. Thus, for practical use, a criterion has been defined for allocating resources for implementation of measures to prevent and counter risks.

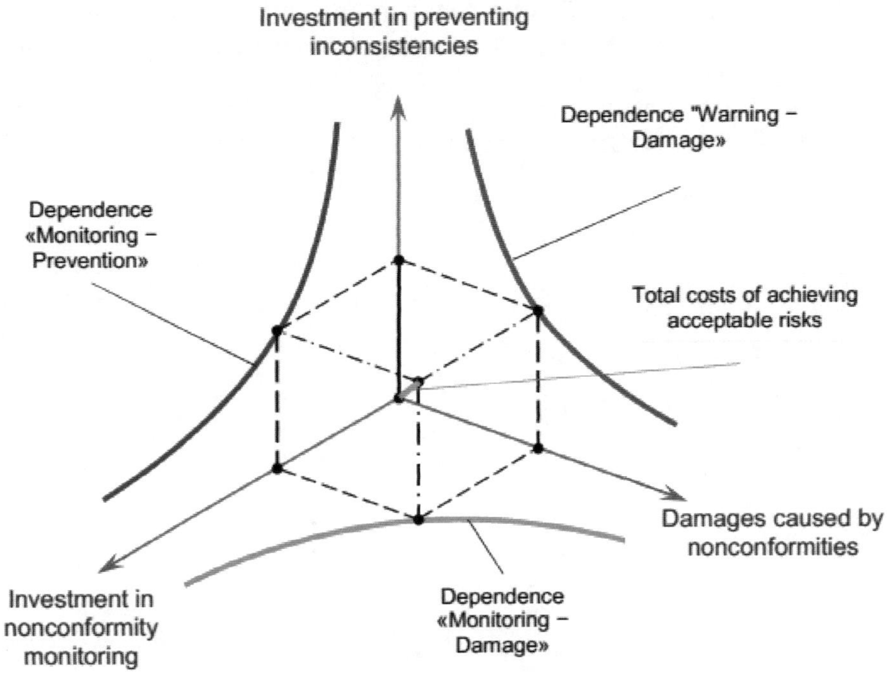

Figure 4. Cost optimization model to achieve acceptable risks.

In addition to event risk analyses (when using the FMEA methodology) [1, 2, 4-10], an appropriate methodological support has been proposed [7] to take into account the risks of variations and bifurcations, both to achieve conformity in quality [9, 10], and when analysing various aspects of risks (bribery [4], occupational safety [5] and energy efficiency [6]) in the supply chain [8].

The presented graphical models allow creating and improving international and state standards - descriptive risk management models that are widely used in practice for the effective and efficient functioning of socio-technical systems [7-10] for various sectors of the economy (mechanical engineering, defense production, etc.).

References

[1] Abrosimov N V *et al* 2015 *Safety of Russia. Legal, Social, Economic, Scientific and Technological Aspects. Scientific Basis of Manmade Safety* ed N A Makutov (Moscow: Znanie publ) p 936

[2] Abrosimov N V *et al* 2018 *Safety of Russia. Legal, Social, Economic, Scientific and Technological Aspects. Manmade, Technological and Aanthropogenic Safety* ed N A Makutov (Moscow: Znanie publ) p 1016

[3] Panov A N *et al* 2009 *Scientific and methodological bases of design. System support of acceptable risks in automotive and agricultural machinery engineering: monograph* (Minsk: BGATU) p 482

[4] ISO 37001-2020 Anti-bribery management systems – Requirements with guidance for use

[5] ISO 45001-2020 Occupational health and safety management systems Requirement with guidance for use

[6] ISO 50001-2018 Energy management systems — Requirements with guidance for use
[7] STB 1506-2015 Management systems Risk management Failure mode and effects analysis method
[8] STB 2484-2016 Management systems Management Verification and validation in the supply chain
[9] STB 16949-2018 Quality management systems Particular requirements for the application of STB ISO 9001-2015 for the organizations involved in the supply chains of Automobile and Tractor, agricultural, material handling, career and special machine building
[10] STB V 15.004-2009 System for development and production of defense products Military equipment Quality management system Requirements Enter. 01.01.2011 (Minsk: State Standard)

Calculated and experimental diagnostics of the stress state of large-sized parts of the chemical industry machines and estimation of their longevity

I M Petrova[1] and M A Filimonov[2]

[1]Mechanical Engineering Research Institute of the Russian Academy of Sciences, Moscow, Russia
[2]JSC RIC "Technoprogress", Moscow, Russia

impetr@mail.ru

Abstract. A series of model and full-scale strain-gauge studies of the stress-strain state (SSS) of a large compressor connecting rod has been carried out. A computational model for assessing SSS is proposed, and its satisfactory agreement with the obtained experimental results is shown. The results of the SSS assessment are used to predict the resource for various operating options.

At a number of enterprises operating nitrogen-hydrogen compressors, there were cases of destruction of connecting rods [1], which creates a dangerous situation. Examination of the destroyed cars showed that all cases of breakdowns are identical and were the result of fatigue cracks in the piston heads. As a rule, the formation of cracks began from the inner surface of the head in its middle section. At the same time, experimental studies of the performance of structural elements that are large in size and work in chemical production are very difficult.

The calculation of the strength of the compressor connecting rod head contains a number of assumptions, it is influenced by the curvature of the head, the angle of sealing, the gap between the finger and the sleeve, and the design features of the heads. In this regard, it is advisable to conduct a study of the stress state of the head by experimental and computational methods using a numerical experiment in order to compare the results obtained. Taking into account that there were no surface damages arising from the action of contact stresses in the examined machines, they were not considered when assessing the probability of failure-free operation of the connecting rod. Previously, the probability of failure of elements of a mechanical system depending on the gaps was analyzed [2] and methods for evaluating mechanical systems by modelling their technical condition were considered.

Method

The method of the research was carried out on the crosshead of the connecting rod of the compressor 6CHBK-355 (piston force of 25 tons). The practice has shown that all of the incident cases of the formation and development of fatigue cracks, destructions of the heads in use were section heads coordinate 80-100^0 from the longitudinal axis of symmetry of the connecting rod (figure 1), indicating the action of maximum stresses in these sections, therefore, the greatest interest is the distribution of stresses in these zones and not at the transition of the head in the beam of the rod.

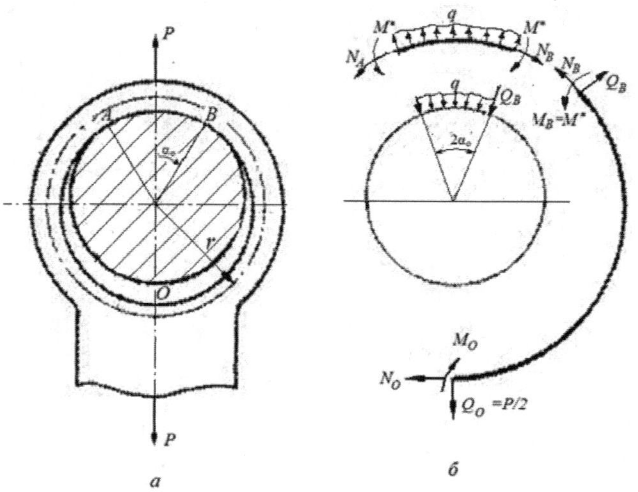

Figure 1. The forces acting in the head of the connecting rod reciprocating compressor.

At the boundary of the contact and gap sections of the pin with the connecting rod bore at points A and B, the radial force NB, tangential force QB, and bending moment MV occur. Since the finger stiffness is several orders of magnitude greater than the rod head stiffness, in accordance with [3], a constant bending moment MV will act on the contact section of the VO. In this case, the curvature of the inner surface of the connecting rod head within the contact angle will be equal to the curvature of the finger. Then on the AB section (figure 1), the curvature is constant and the bending moment is [4]:

$$M^* = \frac{E \cdot I_x(r-\rho^*)}{r\rho^*} \tag{1}$$

where E – elastic modulus, I_x is the moment of inertia, r is the average radius, $\frac{1}{\rho^*} = \frac{1}{r_o+\frac{h}{2}}$ is the curvature of the head in the contact area, r_o- the radius of the gudgeon pin, h – is the height of the section of the head.

From the condition of conjugation of sections, the bending moments are equal to $M_B = M^*$. The moment in the current section, at the beginning of the reference angle α at the border of the sections (contact angle $2\alpha_o$) will be equal to:

$$\mathrm{M} = \mathrm{M}^* + N_\mathrm{B}(1 - cos\alpha) - Q_B r\ sin\alpha \tag{2}$$

Then the equation of the elastic line of the ring in accordance with [3] will have the form:

$$\frac{d^2W}{d\alpha^2} + W = -\frac{Mr}{EI_x} \tag{3}$$

Substituting formula (2) in equation (3) we get

$$\frac{d^2W}{d\alpha^2} + W = -\frac{r^2}{EI_x}[M^* + N_B r(1 - -cos\alpha) - Q_B r\ sin\alpha] \tag{4}$$

where W- is the radial displacement of the contact point $W = -\frac{dV}{d\alpha}$, and r- is the radius of the elastic line of the ring.

The angle of rotation of the normal ϑ is determined from the expression [3]

$$\vartheta = \frac{V}{r} - \frac{1}{r}\frac{dW}{d\alpha}, \qquad (5)$$

where V- is the circumferential offset of the ring point.

For an experimental assessment of the stress state of the connecting rod, full-scale strain-gauge tests were carried out at various operating modes of the compressor (figure 2).

The main tasks of experimental studies of the stress-strain state of the compressor connecting rods were: determining the magnitude and nature of the change in the loads acting on the crank mechanism of the compressor during its operation; determination of the nature of stress distribution in the connecting rod elements with the identification of the most loaded sections; comparison of the results of the numerical experiment with the data of experimental studies. Load cells on the connecting rod head were glued according to the scheme shown in figure 3 (a - inner surface of the connecting rod head; b - outer surface). The necessary measures were taken to protect the load cells from the corrosive environment, temperature influences and electrical interference. To exclude the effect of the arising temperature deformations, compensation strain gauges were used, which were glued to plates made of the same material as the connecting rod. The drying quality of the load cells was checked with a megometer. The places where the strain gauges were glued were covered with a lavsan film, on top of which a sealing coating based on bitumen masses was applied. After applying the coating, the insulation resistance, the integrity of the strain gauges and the absence of a short circuit to the connecting rod weight were additionally monitored. Strain gauge tests were carried out under various operating conditions of the compressor.

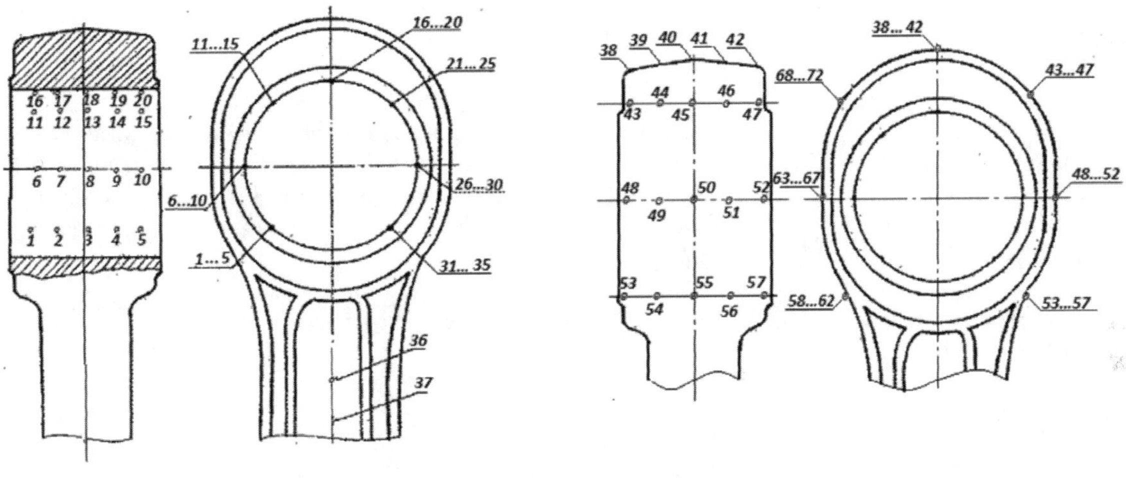

(a) (b)

Figure 2. The location of the strain gauges on the inner (a) and outer (b) surfaces of the connecting rod head.

Main results

As noted earlier in [5], using the boundary conditions for α=0 and $\alpha=\pi-\alpha^0$, we can determine the values of the force factors N and Q as a function of the angle α. The stresses in the zone from the angle α_0 to the angle of the transition of the head to the connecting rod were determined by numerical experiment using formula (4). At the same time, operating loads and gaps in the connecting rod – pin interface varied. The stress state of the connecting rod head was calculated for sections located within the angles from α corresponding to the end of the contact zone to the angle $\varphi = 135^0$. The specified loads corresponded to the operational ones: 180 kN at the nominal operating mode; 250 kN when working with overload and 100 kN when the machine is underloaded. The gaps in the connecting rod pin – connecting rod bore interface were also set in accordance with the operating parameters: the extreme values were 0.05 mm and 0.15 mm. The average gap value was assumed to be 0.1 mm. Based on the results of analytical calculations, which varied the loads in the parts of the crank mechanism and the gaps in the coupling of the connecting rod bore-connecting rod pin, the stresses acting in the most loaded section of the connecting rod head were determined. The stress distribution on the inner surface of the connecting rod head depending on the load and the gap is shown in figure 3.

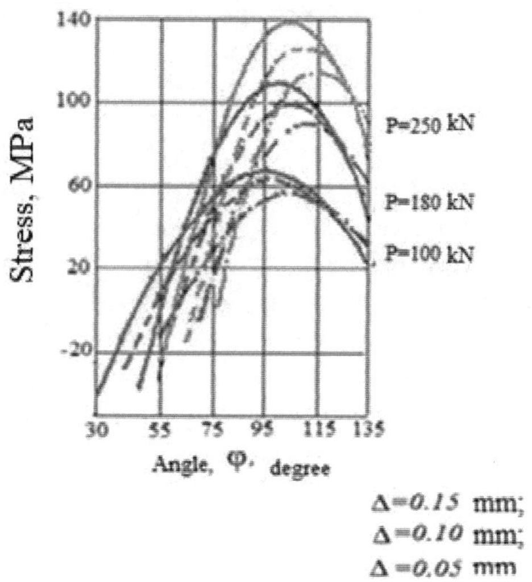

Figure 3. Stress distribution on the inner surface of the connecting rod head obtained by numerical experiment.

Calculation and experimental measurements of stresses showed acceptable agreement.

The characteristics of the connecting rod fatigue resistance are determined according to the recommendations [6, 7] and GOST 25 504 "Calculations and strength tests in mechanical engineering. Methods for calculating fatigue resistance characteristics". The connecting rod is made of 40Cr steel, for which the endurance limit based on reference data can be assumed to be equal to $\sigma - 1 = 280...290$ MPa.

For the fatigue estimation in probabilistic aspect, the methods described in [7] were used. The graph of dependence failure probability on the clearance value is shown in figure 4. It can be seen, that the failure probability is low enough.

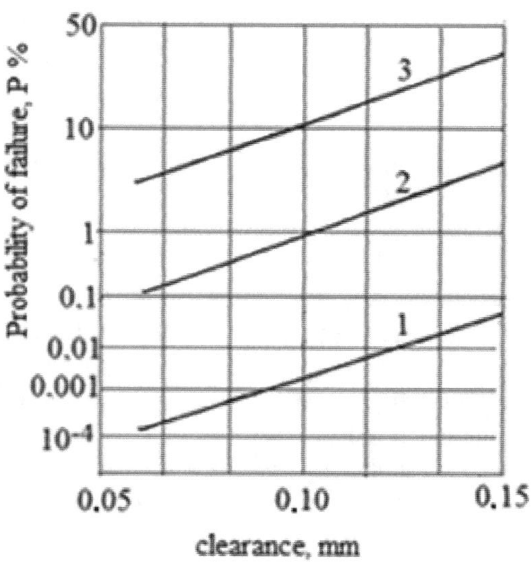

Figure 4. Probability of failure of the connecting rod depending on the load and clearance (Values of piston forces: 1-250 kN; 2 - 180 kN; 3-100 kN).

IOP Conf. Series: Materials Science and Engineering 1023 (2021) 012020 doi:10.1088/1757-899X/1023/1/012020

Conclusions

The numerical experiment allows us to estimate the simultaneous effect of increase and gaps on stress growth, which is practically impossible to achieve by experimental studies. Experimental tests made it possible to determine the stresses in the connecting rod head during operation.

The stresses in the connecting rod head of an oppositional compressor determined by numerical research methods exceed the stresses determined experimentally by 7...22% and give a more conservative estimate of the stress state. The use of computational methods for determining stresses allows us to assess the impact of loads, gaps in the coupling of the pin-connecting rod head and take the necessary measures in advance to maintain the required level of probability of failure-free operation [6, 7].

References

[1] Gusev B M, Nesvizh F A, Petrova I M and Filimonov M A 1986 Improving the reliability of oppositional compressors *Express information. Domestic production experience. Series nitrogen industry* (Moscow: Niithim) **4** 8-10

[2] Petrova I M and Filimonov M A 2018 Probability of failure of elements of the mechanical system depending on the gaps *Problems of mechanical engineering and machine reliability* **2** 25-29

[3] Boyarshinov S V 1973 *Fundamentals of construction mechanics of machines* (Moscow: Mechanical engineering) 456

[4] Golbats-Kokin E M 1969 Calculation of the piston head of the connecting rod taking into account its rigidity and the landing gap in it Energomashinostroenie **8** 39-41

[5] Petrova I M and Filimonov M A 2018 Evaluation of the reliability of compressor connecting rods *Problems of mechanical engineering and machine reliability* **2** 25-29

[6] Kogaev V P 1993 Calculations for strength under stresses, variables in time (Moscow: Mechanical engineering) 363

[7] Calculations of machine parts for endurance in the probabilistic aspect 1991 *Suplimental materials ICSTI 85* ed A P Gusenkov *et al* (Moscow: IMASH RAS)

On the issue of improving the reliability of machines for open-pit mining operations with expiring or expired standard exploitation period

P A Pobegailo [1], A N Ilyina [2] and D Y Kritckij [3]

[1]Mechanical Engineering Research Institute of the Russian Academy of Sciences (IMASH RAN), Moscow, Russia
[2] Federal State-Funded Educational Institution of Higher Education "Moscow Aviation Institute (National Research University)", Moscow, Russia
[3] JSC SUEK-Krasnoyarsk, Krasnoyarsk, Russia

petrp214@yandex.ru

Abstract. The present paper presents our new results related to the research aimed at improving reliability and safety, and reducing operating costs of open-pit mining machines. The main attention was paid by us to the failures of mining equipment. Our main focus has been on machines with the expiring or already expired standard exploitation period, which, however, are not utilized, but continue their work. The failures of mining machines have been collected by us at a number of enterprises in Russia over the past few years. After that, we analyzed and processed them. During the analysis, these initial data were presented by us in the form of frequency graphs. Further, these discrete graphs were "transferred" by us into continuous by means of harmonic analysis methods - using the Fourier series and mathematical package MATLAB. According to the results of the work, it was found out that the number of failures is growing. This can be explained by the rhythm of the nature of the occurrence of failures and connection between the failures and the physiological state of the people working and operating these machines, which was previously not given due attention.

1. Introduction

Mining machines for open-pit mining play an increasing role in the life of mankind, since their effective work provides the planet with the necessary raw materials for survival. The importance of the social and economic role of these machines, the growing demand for them, dictate more and more stringent requirements for their reliability. The reliability of mining machines depends significantly on the quality of work of engineering departments of their operating organizations. The quality of the work of these engineering services can be assessed through the concept of failure. The minimization of their number, the reduction of time and financial resources for their elimination are the main requirements for the work of engineers-operators. At the same time, we cannot affirm that the problem of failures in our country has been completely solved, because of this fact there are known problems with the reliability of mining machines. Some attempts to improve the situation with failures are presented in the present text.

At the same time, we should note that the main emphasis was made on the analysis of failures of mining machines with the expired or expiring normative exploitation period, which is due to the fact that there are a lot of such machines in the park of domestic equipment now. And this problem is new, because in the Soviet times this fact didn't take place.

2. Degree of development of the topic

Serious research on the reliability of mining machines started from the late 50s - early 60s of the last century. At the time the corresponding fundamental works appeared. The number of studies increased until the end of the 90s of the last century and had the widest area of application in all areas of the national economy.

The following engineers and scientists were engaged in research on the reliability of mining machines (and various excavators): Belenkii D.M., Volkov D.P. [1], Getopanov V.N., Demin A.A., Doronin S.V., Kovalchuk V.A., Kravchenko V.M. [2], Krasnikov IU.D. [3], Krylov E.S., Kokh P.I., Makhno D.E., Morozov V.I., Moskvichev V.V., Novikov B.A., Olenev V.A., Pereponov V.I., Petrov I.V., Rakhutin G.S. [4], Reish A.K. [5], Rovkakh S.E., Semencha P.V., Solod S.V.[6], Trop A.E., Fedorov D.I. [7], KHazov B.F. [8 9], SHenderov A.I. [10] and many others.

In addition, it should be noted that many studies and aspects of the reliability problem have been reflected in numerous regulatory and calculation documents, most of which are still relevant today.

However, despite the very significant number of publications and the huge physical and financial efforts to improve the reliability of Soviet and Russian mining machines, this task cannot be said to have been realized. Rather, on the contrary, in recent decades the problem has only grown and appears more and more new faces and colors. And, yes, our mining machines are also inferior in this part to their foreign competitors.

Therefore, any work that brings positive to this area is relevant and in demand.

3. Results

At the beginning of the work in the presented article, its authors collected statistical data on failures on several large (and in something typical) domestic quarries. It should be taken for consideration the data have been collected over a long period of several years. In addition, the authors managed to find some data on failures from these quarries during the Soviet times.

After that, the obtained data were subjected to mathematical processing using the MATLAB package (having created the required program text).

Before using the specified software, we analyzed the distribution of failures according to the time of their occurrence per day, and also studied the question of what time it takes to eliminate certain failures (all this was carried out for the annual interval). In other words, we prepared the initial data for analysis, presenting them in the form we needed, as shown in figure 1. It shows one of the frequency diagrams we received on the distribution of failures by daytime - for the night shift on one of the quarries.

During this phase of the work, we have found that in some cases there is a tendency to increase the number of accidents, decrease productivity and increase costs. Moreover, this is especially evident in the analysis of the condition of bearing metal structures of quarry excavators with the expiring or already expired standard exploitation period.

The essence of the stage of processing this data in the above mathematical package, is to go from discreteness to continuity in our studies, replacing graphs like those shown in figure 1 with graphs like those shown in figures 2 and 3 below. The course of our actions is as follows (as an example based on the frequency graph in figure 1).

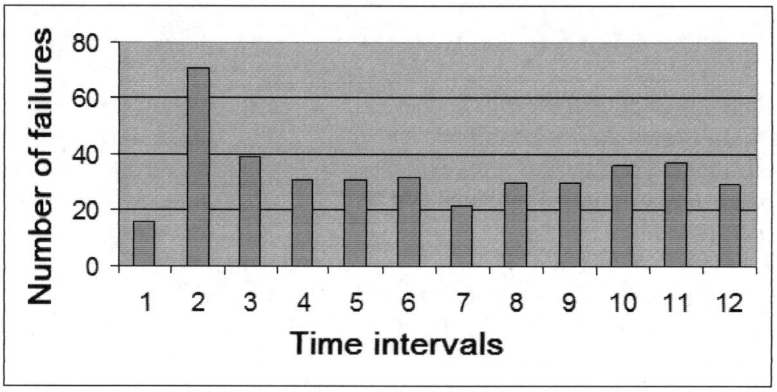

Figure 1. Frequency diagram for night shift [a].

[a] Note: the digit "1" along the abscissa axis is the time interval from 7 to 8 p.m., the digit "2" is the time interval from 8 to 9 p.m., etc. Along the ordinate axis, the number of failures per hour interval (twelve-hour shifts) is marked.

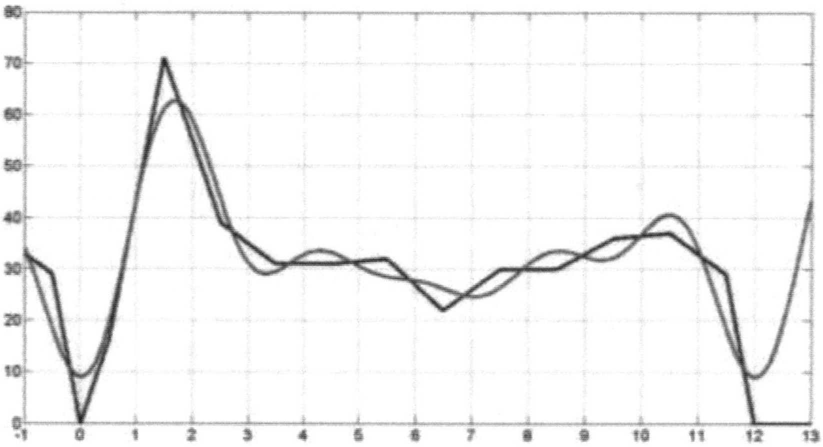

Figure 2. The sum N of the terms of the series at $N = 10$.

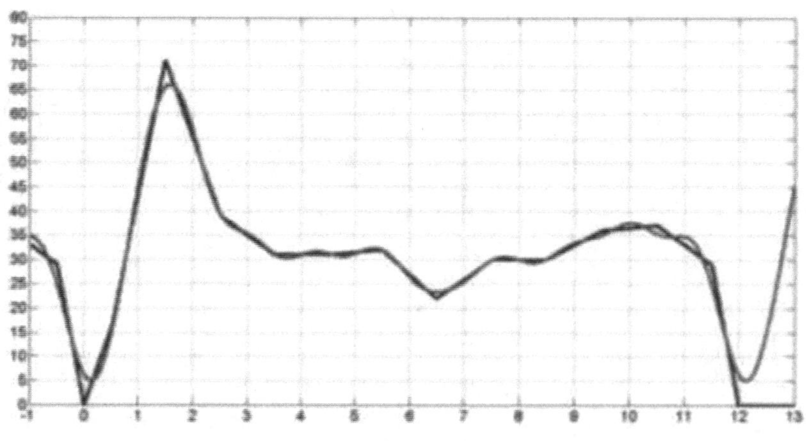

Figure 3. The sum N of the terms of the series at $N = 20$.

First, we build a frequency polygon $f(x)$ that is a broken line connecting the upper bounds of the rectangles that form the frequency diagram.

Next, we process the obtained piecemeal-smooth function $f(x)$ and, using the tools of the MATLAB, we get its representation in the form of the sum of harmonics:

$$f(x)=\frac{1}{2}\cdot 65{,}4583+\left(1{,}0362\cdot 10^{-15}\cdot\cos\left(\frac{\pi x}{12}\right)+3{,}227\cdot 10^{-14}\cdot\sin\left(\frac{\pi x}{12}\right)\right)+$$
$$+\left(2{,}7234\cdot\cos\left(\frac{2\pi x}{12}\right)+4{,}5343\cdot\sin\left(\frac{2\pi x}{12}\right)\right)+ \quad (1)$$
$$+\left(-2{,}8126\cdot 10^{-15}\cdot\cos\left(\frac{3\pi x}{12}\right)+5{,}3291\cdot 10^{-15}\cdot\sin\left(\frac{3\pi x}{12}\right)\right)+\ldots$$

After that, we receive the graphs of the sums of the first 10 and 20 terms of the obtained series (they are shown in figure 2 and figure 3). Let us note that the number of harmonics is now taken by us as an example and it is not a dogma, in other cases there may be a different number of them.

In the regarded example, for the situation with twenty harmonics, the relative error is 0.3% and it seems quite satisfactory for the considered class of problems (the formula for calculating the error is not given now).

When processing the initial data collected by us in the quarries, we have built not only graphs of the type shown in figure 1, but also several more - one of them is a graph of the frequency distribution of the time required to eliminate a single failure (figure 4). These graphs are also subjected to the analysis described above using the approach we have developed.

The result of processing this graph as described above is shown in figure 5.

It should be noted that the processing of the data presented in figure 4 gives the following representation in the form of a sum of harmonics:

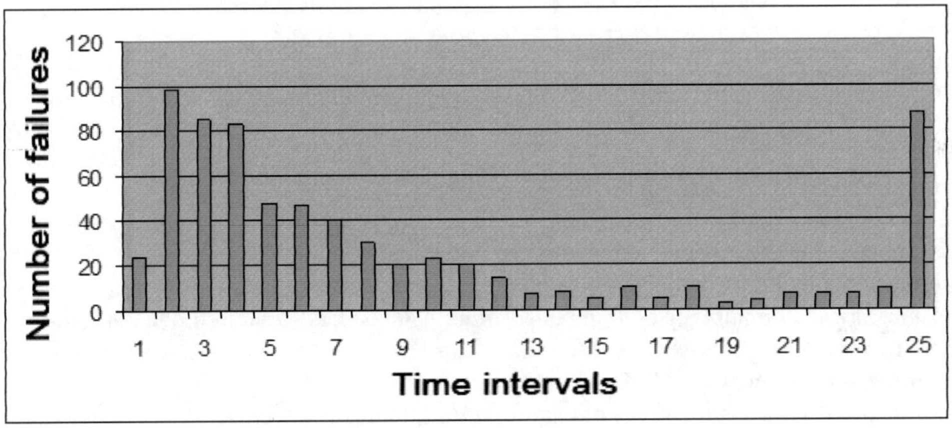

Figure 4. Frequency distribution of time for elimination of single failure[a].

[a] Note: the digit "1" along the abscissa axis is the time interval from 0 to 1 hour for elimination of a failure, the digit "2" is the time interval from 1 to 2 hours for elimination of a failure, etc. (the right interval indicates all failures with the time of elimination more than a day) The number of failures per hour interval is marked along the ordinate axis.

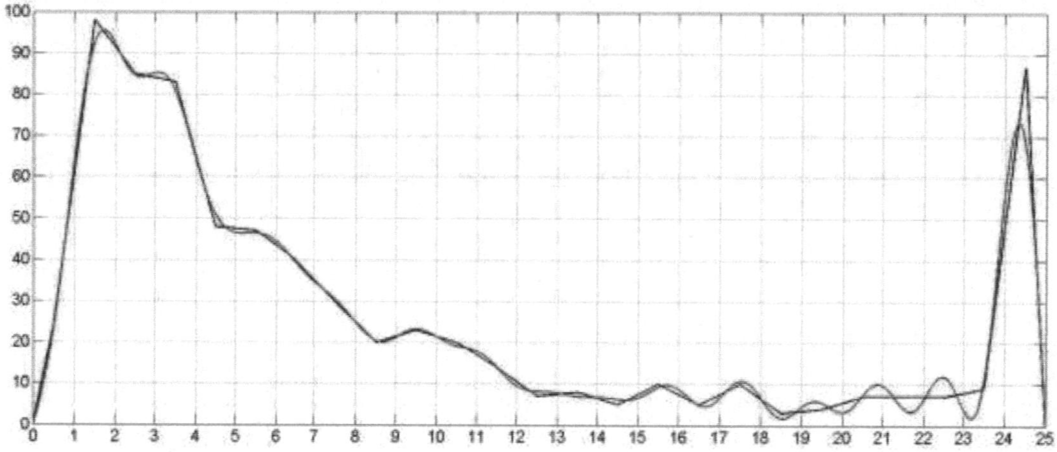

Figure 5. Time of elimination of single failure after processing with harmonic number equal to 30.

$$f(x) = 12{,}9863 \cdot \sin\left(\frac{\pi x}{25}\right) + 11{,}7317 \cdot \sin\left(\frac{2\pi x}{25}\right) + 12{,}5488 \cdot \sin\left(\frac{3\pi x}{25}\right) +$$
$$+ 7{,}1679 \cdot \sin\left(\frac{4\pi x}{25}\right) + \tag{2}$$
$$+ 8{,}2908 \cdot \sin\left(\frac{5\pi x}{25}\right) + 3{,}9859 \cdot \sin\left(\frac{6\pi x}{25}\right) + 6{,}2015 \cdot \sin\left(\frac{7\pi x}{25}\right) + 0{,}9607 \cdot \sin\left(\frac{8\pi x}{25}\right) + \ldots$$

In this case, for the situation with thirty harmonics, the relative error is 0.3%.

4. Conclusion

In this work, for the first time in our country, the authors formulated the most important and still little-solved problem of reducing the number of failures during the operation of mining machines in open mining work that have the expiring or expired standard exploitation period (and thereby reducing costs and ensuring productivity required in the mines).

At the same time, on the basis of the significant collected amount of initial information, the important applied results were obtained (not given in the text).

The major theoretical results of this work are:

- an increase in the number of failures of mining machines in some domestic quarries located either in front of Rubikon or already after it in terms of their normative standard exploitation period was revealed;
- on one of the quarries the rhythmic nature of failure occurrence is revealed that is qualitatively the same for day and night shifts. At night, the number of failures is greater (seasonal rhythms are qualitatively the same as in the Soviet times). It provides for the necessity for further research into the impact of human biological features on the operation of mining machines;
- frequency graphs can be successfully processed by means of harmonic analysis.

At the end of this text, we note that the introduction into the practice of operating mining machines of the approach proposed by the authors of this article to the analysis of failures will increase the reliability of mining machines, reduce their accident rate and financial costs for their repair.

We also note that this study is included both in the system of express diagnostics of quarry excavators created by its authors and closely concerns the methodology for designing hydraulic-driven quarry excavators that we are developing.

References

[1] Volkov D P, Nikolaev S N and Marchenko I A 1972 *Reliability of rotary trench excavators* (Moscow: Mashinostroenie) p 207

[2] Kravchenko V M and Rusikhin V I 2002 *Repair technology of mechanical shovels* (Moscow: MGGU Publishing House) p 231

[3] Krasnikov IU D, Solod S V and Khazanov KH I 1989 *Improving the reliability of mining dredging machines* (Moscow: Nedra) p 215

[4] Rakhutin G S 1970 *Probabilistic methods for calculating reliability, prevention and reserve of mining machines* (Moscow: Nauka) p 203

[5] Reish A K 1986 *Increasing the wear resistance of construction and road machines* (Moscow: Mashinostroenie) p 184

[6] Solod S V 2003 *Reliability of mining dredging machines* (Moscow: OOO "Nedra-Biznestsentr") p 291

[7] Fedorov D I and Bondarovich B A 1981 *Reliability of working equipment of earth-moving machines* (Moscow: Mashinostroenie) p 280

[8] KHazov B F 1979 *Reliability of construction and road machines* (Moscow: Mashinostroenie) p 192

[9] KHazov B F and Didusev B A 1986 *Handbook on machine reliability calculation at the design stage* (Moscow: Mashinostroenie) p 224

[10] SHenderov A I, Emelianov O A and Odin I M 1976 *Reliability of mining transport equipment* (Moscow: Nedra) p 247

Technological biomechanics of fibrous composites

A N Polilov and N A Tatus'

Mechanical Engineering Research Institute of the Russian Academy of Sciences, 4 Maly Kharitonyevsky Pereulok, Moscow, 101990, Russia

polilovan@mail.ru

Abstract. There are numerous examples of using Natural solutions in technology, especially, in design of composite structures [1]. Bio-inspired methods (suggested by Nature) are widely used in human activity. For instance: how does branch-trunk connection work? You can break a branch, but nearly impossible to tear the branch off the trunk. The connection turns out to be stronger than the part itself. So far, designers of composite products can only dream of such methods of joining.

1. Methods for constructing biosimilar curvilinear fiber trajectories

Nature never uses right linear reinforcement. Nature observation led to the idea of the need to use an optimal, curved reinforcement structure in composite design, when the fiber paths are consistent with the stress fields. A large number of works [2, 3, etc.] are devoted to the construction of optimal trajectories of fibers "flowing around" a circular hole.

In [3] the new approach to the construction of fiber paths coinciding with the trajectories of the main tensile stresses is developed. Iterative continuous fiber tracing makes it possible to go directly to additive technologies that provide a rational structure of curvilinear reinforcement in the area of holes or attachment points. The fibers trajectories flowing around the hole resemble the structure of wood in the knot zone (figure 1).

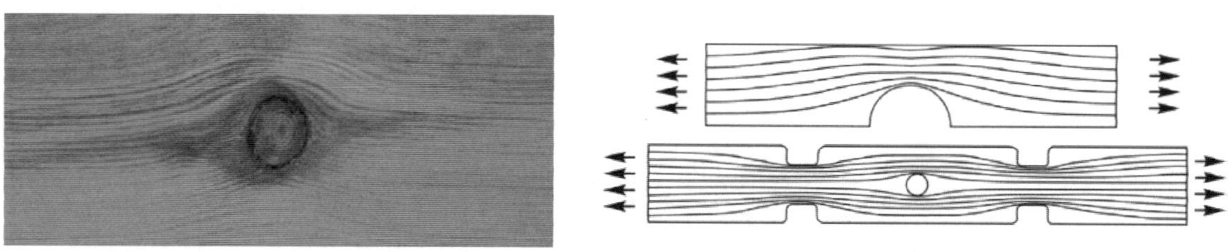

(a) (b)

Figure 1. (a) structure of pine wood in the knot zone, (b) uniformly stressed structure of fibers laying along the trajectories of the maximum principal stresses near holes and undercuts in a plate under tension.

The main conclusion from the analysis is that the maximum stress "per fiber" with an optimal reinforcement structure is about 3-4 times less than with a homogeneous rectilinear laying, that is, the effective stress concentration factor decreases from about 5 to 1.3, and the fiber overload at the bottom of the hole is only about 30%. At the same time, it is extremely important that the shear stresses that cause splitting near the holes disappear.

It is obvious that the harmful effect of stress concentration at the composite parts joints can be reduced precisely by using a special reinforcing structure, consistent with stress field.

2. Biomechanical design principles for composite parts joining

One of the fundamental tasks of composite design is the development of joining methods that achieve high fiber strength.

The fundamental disadvantages of all known joining methods force us to study Nature experience, and the knot "design" can suggest the optimal fibers trajectories in the composite joining with a bolt or rivet (figure 2).

Rational reinforcement leads to the significant reduction in local stresses per fiber, elimination of splitting and the increase (by at least 50%) in the bearing capacity of the joint.

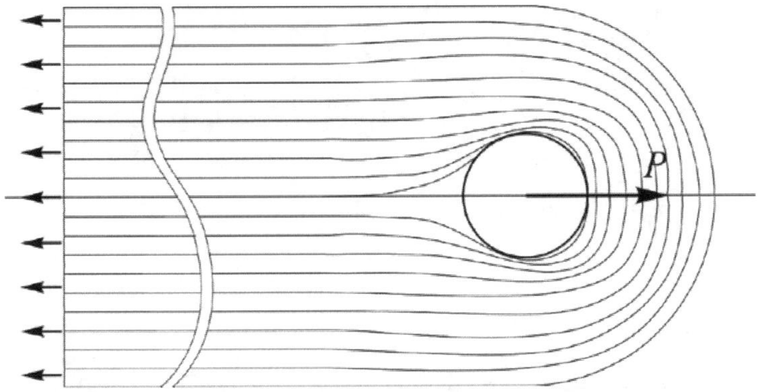

Figure 2. Rational trajectories of fibers envelop a loaded hole.

3. Design of bioinspired branching and profiled composite elastic elements

The secrets of the tree crown structure have attracted the attention of researchers for many centuries. Even Leonardo da Vinci in his notes made the following statement: "The sum of the squares of the branches diameters is the same before and after branching". The "Leonardo's Rule" suggests the way to create branched (figure 3) and profiled (figure 4) composite elastic elements [4, 5].

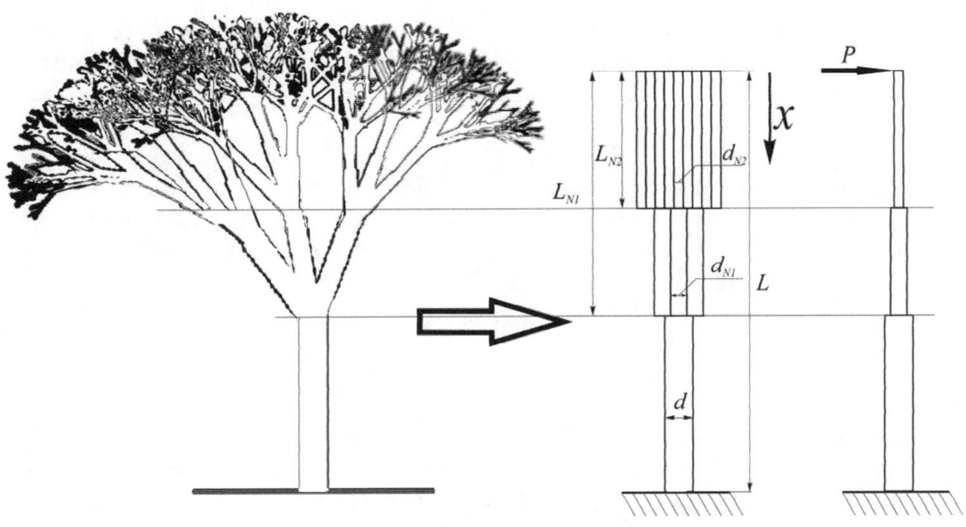

Figure 3. Scheme of equal-strong branching of a circular cross-section rod.

When loading a perfectly branched or profiled beam with a variable bending moment:

$$M(\bar{x}) = M(1)\bar{x}^{\gamma};$$
$$\bar{x} = \frac{x}{l}. \qquad (1)$$

to maximize the stored elastic energy, sequential branching into an integer number of branches is advantageous $1+2\gamma$, and the maximum possible mass reduction factor is also equal $1+2\gamma$. Let`s note that loading by the end force P corresponds $\gamma=1$ in (1), $\gamma=2$ corresponds to a uniformly distributed load, $\gamma=3$ corresponds to a linearly varying load.

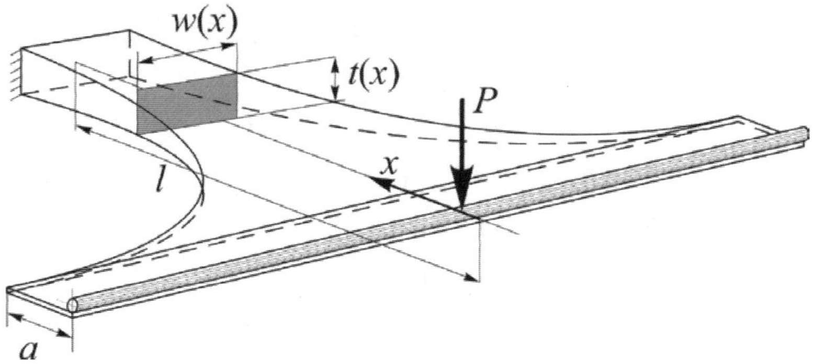

Figure 4. . Constant-area beam with constant cross-sectional area.

Branching uniformly stressed composite elastic elements are as effective as profiled constant-area beams. The advantage of branching over profiling consists in conservation continuous rectilinear reinforcement and in the possibility of dimensions reducing when collecting "branches" into a bundle.

4. Additive biomimetic technologies

The development of additive 3D printing technologies expands the possibilities for creating optimal designs with curved fiber trajectories. On this way, it is possible not only to design and create products of various shapes, but also to manufacture bio-inspired joints [6], which are much more efficient than traditional "metal-like" joints.

Since with such technologies there is no need for high temperatures and significant molding efforts, in future it is possible to create technological sections for manufacturing and repairing composite elements directly in orbit [7].

Conclusion

Let us point out the three most promising, in our opinion, directions in the composites mechanics: strength biomechanics, computer simulation of optimal structures, and composites technological mechanics. It is the modeling of biotechnology and biomaterials that should provide a new breakthrough in the creation of composite structures.

References

[1] Polilov A N and Tatus' N A 2018 Strength biomechanics of fibrous composites (Moscow: FIZMATLIT) p 328
[2] Papapetrou V S, Patel C and Tamijani A Y 2020 Stiffness-based optimization framework for the topology and fiber paths of continuous fiber composites *Compos. Part B-Eng* 183 107681
[3] Malakhov A V and Polilov A N 2016 Design of composite structures reinforced curvilinear fibres using FEM *Compos. Part A* vol 87 pp 323–28

[4] Polilov A N, Tatus' N A, Tian X and Arutjunova A S 2019 Equistrong branchy composite beams with a constant total area of variable elliptic cross sections *Mechanics of Composite Materials* vol 55 **3** pp 465—82
[5] Polilov A N, Tatus' N A and Tian X 2019 Analyzing of the correctness of equal strength composite profiled beam bending problems *J. of Applied Mech. and Tech. Physics* vol 60 **1** pp 144–55
[6] Malakhov A V, Polilov A N, Zhang J, Hou Z and Tian X 2020 A modeling method of continuous fiber paths for additive manufacturing (3D printing) of variable stiffness composite structures *Applied Composite Materials* vol 27 pp 185-208
[7] Misochenko A A, Prozhega M V, Salamandra K B, Tatus' N A and Shokhin A E 2015 Orbital technological complex (ORTeK) for the composite structures production in outer space *In the book: Vostochny Cosmodrome and Prospects for the Development of Russian Cosmonautics* pp 64-65

Acknowledgments
This work was supported by the Russian Foundation for Basic Research (grant 18-08-00372-A).

IOP Conf. Series: Materials Science and Engineering 1023 (2021) 012023 doi:10.1088/1757-899X/1023/1/012023

Ceramic coatings synthesis on valve metals and alloys by micro-arc oxidation

D Rakov

Mechanical Engineering Research Institute of the Russian Academy of Sciences (IMASH), Moscow, Russia

rdl@mail.ru

Abstract. The most pressing tasks in engineering and materials science are the synthesis of high-quality coatings with high durability and corrosion resistance. This paper presents ceramic coatings synthesis on valve metals and alloys by micro-arc oxidation. The synthesys based on advanced morphological approach. The approach is to generate a sequence of algorithms, comparison and selection of the set of rational variants of technological solutions. The approach allow to reduce the dimensions of the morphological set and improve the efficiency of search procedures.

1. Introduction

The most pressing tasks in engineering and materials science are the synthesis of high-quality coatings with high durability and corrosion resistance. The search is under way for new technologies to provide multifunctional coatings. Micro-arc (plasma) oxidation (MAO) consists in the formation of a part on the surface under the influence of micro-arc discharges of high-strength wear-resistant coating, consisting mainly of Al_2O_3 (corund) and other aluminium oxides. The complexity of the flowing physical and chemical processes leads to consideration of various options for the implementation of MAO-coating [1-4].

The investigated set of technological processes is relatively new and rapidly developing. Therefore, the potential number of technology variants is large. As a consequence, morphological approaches can be used for classification and synthesis of technologies.

2. Morphological approach

For the stage of structural analysis and synthesis of innovative engineering and technological systems (TS) the intuitive and discursive methods can be used [5, 6]. The most common method among the discursive techniques is morphological analysis with building of morphological matrix (MM) (figure 1). By frequency of use, morphological methods are the first among ranks of discursive approaches. Thus, according to statistics compiled in 2009 by German scientists, the total number of firms using the morphology is more than 40%, while regularly use it more than 20% [7]. Morphological synthesis is regarded as a methodology to streamline the problem to be solved. At present, there are many methods to search and synthesize solutions based on the morphological analysis in in a variety of physical and engineering areas [8, 9]. One solution of this problem, in our view, consists in using models of the clustering and set theory. In this respect, it seems appropriate to resort to advanced morphological approach (AMA) with following physical-mathematical modeling and experiments [10, 11].

Options →

Attributes ↓

	P^1	P^2	P^3	...	P^k
P_1	P_1^1	P_1^2	P_1^3	...	P_1^k
P_2	P_2^1	P_2^2	P_2^3	...	P_2^k
P_3	P_3^1	P_3^2	P_3^3	...	P_3^k
...	...	...	...	...	...
P_n	P_n^1	P_n^2	P_n^3	...	P_n^k

Figure 1. Morphological matrix.

3. Micro-arc oxidation technology synthesis

The main design and technological parameters influencing the synthesis of MAO coatings include the used modes of coating formation

- by changing electrical parameters - potentiomeric or galvanostatic;
- on the polarity of the applied voltage (anode, cathode, anode-cathode, cycling regimes of different polarity with pauses);
- ratios of cathode and anode modes, as well as pauses between them;
- used electrolytes by number of components and chemical nature - silicate-alkaline, sour, etc.;
- on frequency control of processes and possible elemental base of process management;
- by the way of forming a discharge (soft, soft-hard, hard-soft, hard) etc.

The main features that provide the process of MAO include the main modes of coating formation, divided by [1-3]:

1. Energy parameters:
   - changes in electrical values (potentiomeric or galvanostatic);
   - the polarity of the applied voltage (anode, cathode, anode-cathode, cycling modes of different The ratios of cathode and anode modes, as well as pauses between them;
   - the way of forming a discharge (soft, soft-hard, hard-soft, hard) etc.
2. Used electrolytes:
   - using a number of components;
   - chemical nature (silicate-alkaline, acidic, etc.);
3. Control of application processes:
   - frequency process control and possible elemental base;
   - process control.
4. Material for application - vent metals and alloys, graphite, etc.

To assess the options with the help of expert assessments, MM was drawn up with the following features - coating modes, electrolytes used, materials used, application control, frequency regulation, etc. Any set of elements of all traits (one of each line of the morphological table) is a possible version of the technology. The power of the morphological set is 20736 variants [3].

Twelve TS reference variants, formed from patent descriptions and well-known solutions, were also included in MM. In the future, a number of criteria were chosen, characterizing the constructive

IOP Conf. Series: Materials Science and Engineering 1023 (2021) 012023 doi:10.1088/1757-899X/1023/1/012023

and economic indicators of TS and expertly assigned weights on the basis of the method of attributing points, which has an advantage in choosing the freedom of action over the methods of ranking.

During synthesys 1,648 variants were generated and 256 selected for further clustering. Of the 256 generated and selected variants, 9 clusters were formed (figures 2, 3). Among all clusters, cluster 9 has the highest rating (by 14% than the average estimate), which was primarily investigated.

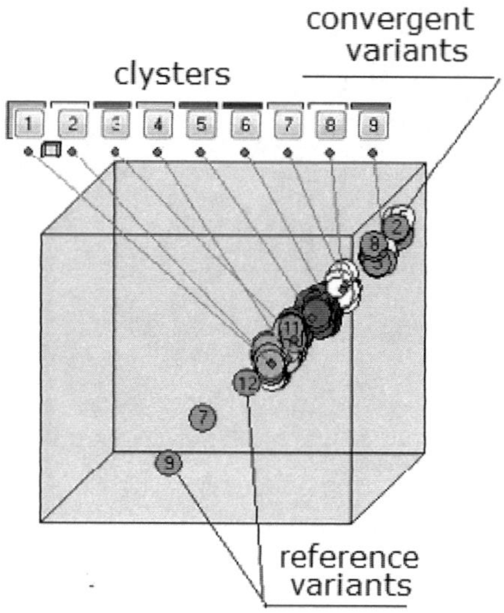

Figure 2. Morphological solutions space.

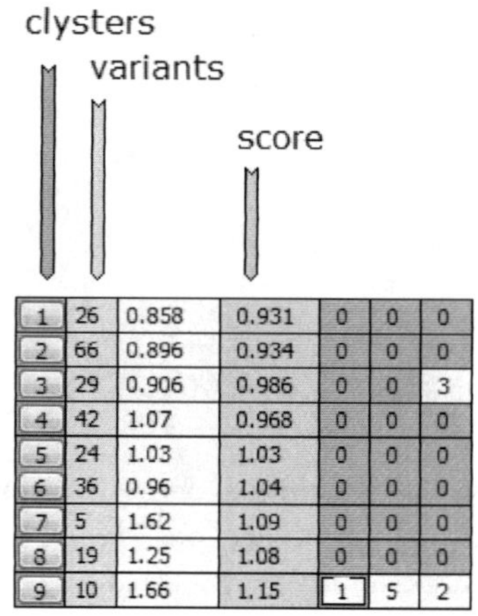

1	26	0.858	0.931	0	0	0
2	66	0.896	0.934	0	0	0
3	29	0.906	0.986	0	0	3
4	42	1.07	0.968	0	0	0
5	24	1.03	1.03	0	0	0
6	36	0.96	1.04	0	0	0
7	5	1.62	1.09	0	0	0
8	19	1.25	1.08	0	0	0
9	10	1.66	1.15	1	5	2

Figure 3. Clusters comparison.

A specific feature of the reference variants and most variants of this cluster is the control of flowing processes with the help of bipolar transistors with an insulated gate bipolar gate transistor (IGBT), which leads to an increase in the ability to control and control. These variants differ from other variants - inventions of higher characteristics [12].

As a result, TS was created, which is a combination of variants 8-10 and extended in the sign of the used electrolytes. TS is characterized by a wide range of process control capabilities, in particular, at the initial stage only anode mode is implemented, which allows to quickly increase the ceramic coating, in the future anode-cathode mode of application with frequency control with the help of IGBT controlled by the microcontroller. The transistor involves anode and cathode voltage on the parts when they reach the level of voltage they need for the process and trims the pulses of the current at the time of the termination of the process. The latency time between cathode and anode cycles is also set. For example, the cathode pulse that prepares the work of the anode should lie within 0.5...1. Compliance with these conditions allows you to optimize the coverage process. In addition, the process is soft, the full use of the useful part of the current pulses, and thus the minimum electricity consumption and high quality of coverage.

The device allows to control technological processes within wide limits (figure 4). Technological process is realized in the following way. Detail 5 is placed in the electrolyte and connected to the anode of power unit 2. Transistor includes anode and cathode voltage on the parts at the moment when it reaches necessary for the process level and cuts off current pulses at the moment of process termination at duration of pulses t_A and t_K (figure 5). The delay time between cathode and anode cycles t_{Ak} is also set. The fulfillment of these conditions allows optimizing the process of coverage. In addition, a soft start of the process, full use of the useful part of current pulses and thus a minimum power consumption and high quality of the coating are provided.

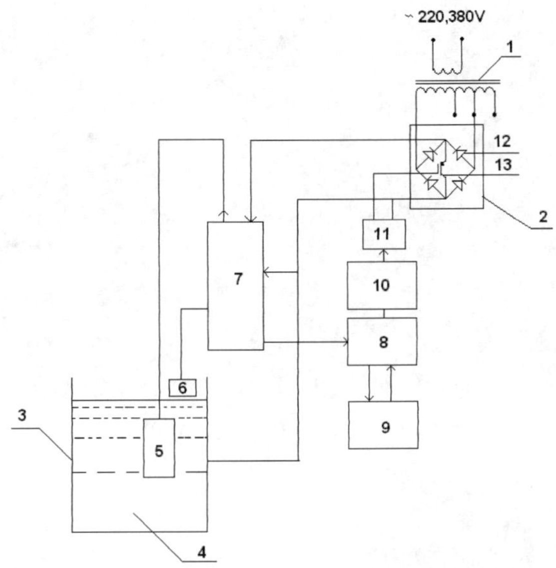

Figure 4. Principal scheme of MAO installation with automated coating control process.

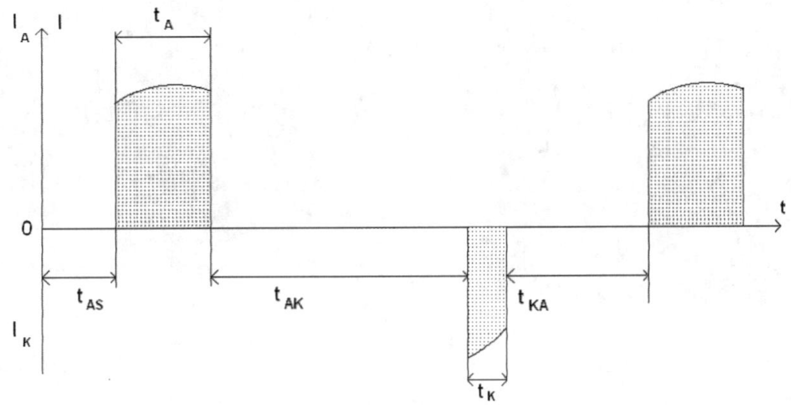

Figure 5. Processing process control (A - anode, K - cathode, IA, IK - anode and cathode electric current, tAK, tKA - temporary delays between the onset of anode and cathode cycles).

The proposed TS makes it possible to fully regulate the application process, so the improvement in the quality of coating is achieved, for example, by the fact that at the beginning of the coating the process is completely in anode mode.

A pilot installation of MAO with extensive process control capabilities has been created for the chosen TS. The installation obtained hard-to-wear ceramic coatings on valve metals and alloys (figures 6-8).

Figure 6. Pilot MAO installation.

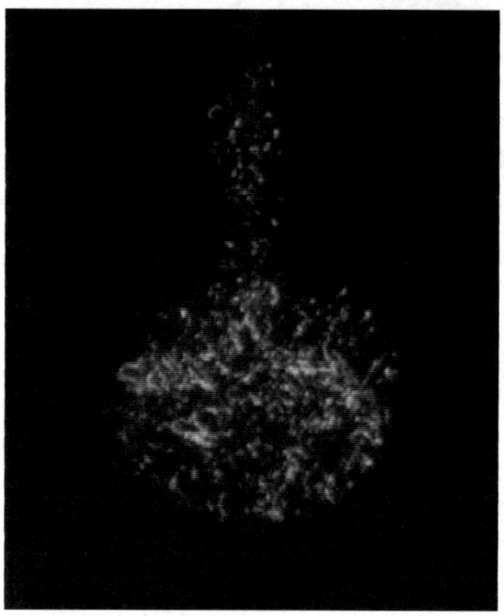

Figure 7. The process of micro-arc oxidation.

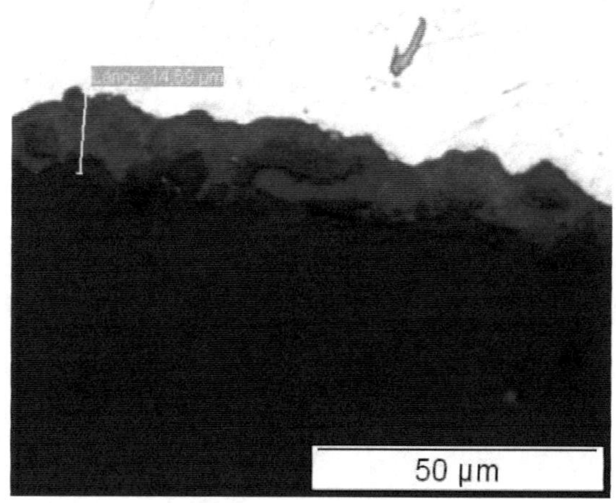

Figure 8. Ceramic coating applied.

4. Conclusion

The proposed approach gives specification and order of structurization of the problem being solved. The degree of substantiation of the decisions made increases. Having analyzed a lot of decisions, it is possible to narrow down the amount of information for making a final decision by a specialist.

For the process of micro-arc oxidation, an experimental setup has been created and the following characteristics have been achieved:

- the possibility of coating on complex products, internal surfaces and hidden cavities;
- obtaining coatings of thickness from 0.05-0.2 mm to 0.3-0.8 mm with adhesion comparable to the strength of the substrate material;
- possibility of full automation of the process of micro-arc oxidation itself;
- wide possibilities of process speed regulation (by current, voltage, frequency characteristics and ratio of anode and cathode components).

References

[1] Suminov I V, Belkin P N, Epelfeld A V, Ludin V B, Crete B L and Borisov A M 2011 *Plasma-electrolytic modification of the surface of metals and alloys* (Moscow: Technosphere)

[2] Rakoch A G, Gladkova A A, Linn Z and Strekalina D M 2015 The evidence of cathodic micro-arc discharges during plasma electrolytic oxidation of kight metallic alloys ans micro-arc discharge inyensity depending on pH electrolyte *Surface and Coatings Technology* **1** 138-44

[3] Klimenko B and Rakov D 2018 Analysis and synthesis of innovative engineering solutions and technologies based on advanced morphological approach *Advances in Artificial Systems for Medicine and Education. AIMEE 2017. Advances in Intelligent Systems and Computing* **658** 274-83

[4] Yerokhin A L, Snizhko L O, Gurevina N L, Leyland A, Pilkington A and Mattthews A 2003 Discharge characterization in plasma electrolytic oxidation of aluminium *J. Phys. D: Appl. Phys*. **36** 2110–20

[5] Zwicky F 1969 *Discovery, Invention, Research – Through the Morphological Approach*, (Toronto: The Macmillan Company)

[6] Ritchey T 2006 Problem Structuring using computer-aided morphological analysis *J. of the Operational Research Society Special Issue on Problem Structuring Methods* **57** 792-801

[7] Smerlinski M, Stephan M and Gundlach C 2009 *Innovationsmanagement in hessischen Unternehmen. Eine empirische Untersuchung zur Praxis in klein- und mittelständischen Unternehmen* (Marburg)

[8] Isenmann R, Landwehr-Zloch S and Zinn S 2020 Morphological box for ESD – landmark for universities implementing education for sustainable development (ESD) *The Int. J. of Management Education*

[9] Arciszewski T 2018 Morphological analysis in inventive engineering *Technological Forecasting and Social Change* **126** 92-101

[10] Rakov D 2011 Structural analysis and synthesis of new engineering systems on morphological approach (Moscow: URSS)

[11] Rakov D 2019 Okkam - Advanced Morphological Approach as Method for Computer Aided Innovation (CAI) *MATEC Web of Conf. 298 ICMTMTE 2019 00120*

[12] Klimenko B, Rakov D and Thorbek J 2011 Device for micro-arc oxidation of valve metals and alloys *Patent Russian Federation* No 2413040

IOP Conf. Series: Materials Science and Engineering 1023 (2021) 012024 doi:10.1088/1757-899X/1023/1/012024

Development of multilevel models for assessment of risk of train derailment

D O Reznikov

Mechanical Engineering Research Institute of the Russian Academy of Sciences, 101990, 4 Maly Kharitoniecsky lane, Moscow, Russia

mibsts@mail.ru

Abstract. The paper presents a multilevel model for assessing risk of train derailment. The model is based on the toolkit of Bayesian nets and allows accounting for the influence of human and organizational factors as well as factors of safety regulation regimes on the occurrence of stochastic events that compile various scenarios of railroad accidents.

1. Introduction

The statistics of major accidents at rail transport facilities indicates that these accidents are, to a large extent, related to errors of employees of the railway company, as well as to organizational and management factors and to factors of safety regulation imposed by supervising agencies. According to existing estimates, the number of accidents that is associated with human and/or organizational factors varies for different sectors of the economy in the range from 50 to 90% of their total number [1]. Such a significant scatter in data is determined by the difference in the interpretation of the concepts of *technical failure* and *failure associated with the human error* since failures of the technical system components at the operational stage is often caused by human errors made ay the stages of their design, construction, maintenance, monitoring, etc. Therefore, if you follow the chain of cause-and-effect relationships far enough, then almost any technical failure can be attributed to a human error in the recent or more distant past.

Issues related to the human factor, in one degree or another, are taken into account when conducting a classical risk analysis of the railway infrastructure and rolling stock facilities. In particular, errors of drivers, dispatchers, or repair workers can be included in failure scenarios as initiating or intermediate events. However, it often happens that human errors only affect the probability of failure of elements of a technical system. If these errors only lead to the weakening of individual components of the system such errors often cannot be explicitly taken into account in the framework of the traditional risk analysis. For example, violations of maintenance procedures often lead to a decrease in the reliability of components of a technical system, but are not a direct cause of their failure.

In addition, the construction of failure scenarios of the technical system does not allow taking into account the fact that human errors, in turn, are often associated with organizational and management problems: for example, an inadequate economic incentives for employees of a railway company, insufficient professional training of personnel, or unreasonable limitation of investments in safety at the company level. It must also be taken into account that organizational and management factors operating within a separate company are formed under the influence of safety regulation regime, which is expressed both in the form of direct regulation of safety measures that are to be implemented and in the form of economic incentives for companies to reduce risk.

When conducting a quantitative risk analysis, one has to face the uncertainty associated with: (1) the limited knowledge about the technical system in normal and emergency conditions, and (2) the variability of the parameters of the technical system and the external environment of the system. In the case of the inclusion of the human factor, the level of uncertainty increases even more, due to the inability to accurately predict the physical and psychological state of the employees (the driver, dispatcher, or repair worker) at various points in time during normal and emergency situations, and, therefore, to predict possible errors. The latter, as a rule, are not independent events, but are caused by the influence of the organizational environment in which the company employees operate. The influence of organization and management factors on the actions of employees also cannot be unambiguously calculated.

In any case, human and organizational factors, as well as factors of transportation safety regulation should be explicitly taken into account when conducting a quantitative analysis of risks generated by the railway infrastructure. Therefore, the model for integrated risk assessment should have four levels and include: (1) a probabilistic analysis of the failure scenarios of the technical system; (2) analysis of decisions and actions of operators that affect basic events in the failure scenarios of a technical system; (3) the study of management factors within the company that affect the decisions and actions of operators; (4) factors of transportation safety regulation, under the influence of which management principles are formed at the company level. Probabilistic relationships between model parameters related to these four levels should be quantified using the available theoretical concepts, statistics, and expert estimates. The idea of development of the multilevel risk assessment models was proposed by E.Pate-Cornell and D.Murthy in [2].

2. Building a multi-level risk management model

The organization of the work and the management regime at the level of the railway company are factors that determine the likelihood of inadequate/unauthorized actions by personnel. Organizational and management factors are risk management levers that include, in particular, staff selection and training mechanisms; organization of monitoring systems for the technical condition of the facility. These control levers can affect the state of an individual employee (the level of fatigue, professional and psychological training, etc.), they can also create an environment that affects decision-making (for example, through material interest, information support for the decision-making process, approved rules and procedures actions in regular and contingency situations, methods for selecting and training personnel, a system of fines for violations of safety regulations and incentives for their observance, budget gender enterprise ethics regarding investing in monitoring and control systems for the technical condition of rolling stock and infrastructure facilities, etc.).

The mechanisms of transportation safety regulation and incentives for the organizations of the Russian Railways holding to ensure safety at the facilities they operate can be considered as top-level factors affecting the safety practices applied within individual companies.

In order to take into account the four-level structure of influences, an integrated model must be developed (figure 1) that includes the assessment of:

- failure scenarios of the technical system (level 1);
- human factor, including the systematic identification and probabilistic description of decisions and actions of operators that affect the probability of the occurrence of basic events of failure scenarios (level 2);
- organizational and management factors at the level of the railroad transportation company that affect the likelihood of making certain decisions and actions (level 3);
- factors of state safety regulation (level 4).

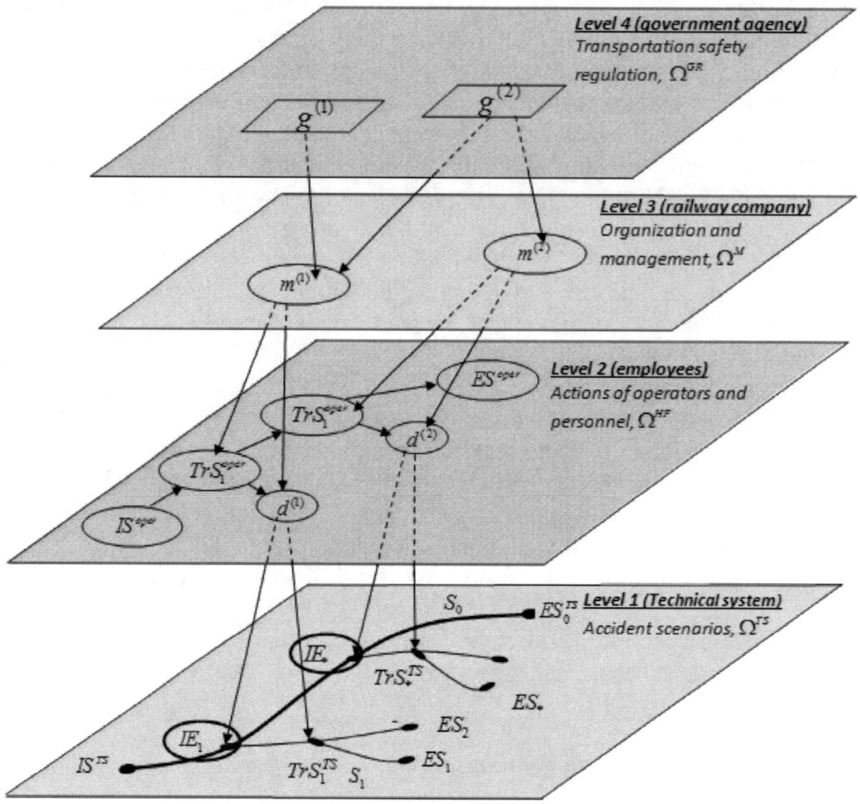

Figure 1. Multilevel model of risk analysis.

The four-level model allows comparing different methods for managing risk generated by the railway infrastructure and the rolling stock: improving the elements of the technical system, reducing the likelihood of errors and unauthorized actions by operators, improving the organization and management within the company, optimizing the state regulation regime. The structure of the relationship between the technical system, the actions of the operator (personnel) and management factors are shown in figure 1.

Analysis of the multi-level model should start from the bottom [2, 3].

Level 1. Technical system

Changes occurring in the technical system designed to ensure the implementation of the transportation process can be presented as a path in the space of the technical system states. The path defines the transition from the initial state (*IS*) of the system (putting it into operation, figure 1) to its end state or decommissioning (ES_0) [4]. When it is possible to ensure such a transition, they say that the system has implemented the specified scenario or success scenario (S_0). If an initiating event occurs (*IE*) in the system (a component failure or extreme external impact), it can deviate from the scenario and proceed to implement new scenarios ending with end states (ES_i) different from the designed end state (ES_0).

In this case, we can say that a failure occurred in the system due to its inability to reach the required end state.

When considering first the technical system itself, according to the total probability theorem, the probability of a system failure can be written as:

$$P(F) = \sum_i P(F \mid IE_i) \cdot P(IE_i) \tag{1}$$

where $P(IE_i)$ is the probability of occurrence of the initiating event, $P(F|IE_i)$ is the conditional probability of failure in the case of the initiating event occurrence.

In the general case multiple failure scenarios leading to different end states (ES_1, ES_2,…, ES_m) can be implemented in the system. Probabilities of occurrence of various end states $P(ES_i)$ can be obtained as:

$$P(ES_h) = \sum_i P(ES_h \mid IE_i) \cdot P(IE_i), \; h = 1, 2, 3, \cdots, m \qquad (2)$$

Level 2. Human Factor

The action (decision) made by the operator at the time of decision-making (t_*) is considered as a random variable (event) that takes one of possible values $\{d_1^{(t_*)}, d_2^{(t_*)}, ..., d_{l_*}^{(t_*)}\}$ (options for the actions of the operator). The choice of an option is determined by the physical and psychological state of the employee at the time of decision-making, the level of his professional qualification and awareness of the real state of the system, the existing rules of action in emergency situations and incentives. Thus, all options for the actions of employees in the process of escalation of an accident are grouped into categories $d^{(t_1)}$, $d^{(t_2)}$,…, $d^{(t_f)}$ corresponding to different points in time of decision-making. Each occurence of the process of escalation of the accident will proceed with a certain set of decisions made by operators $d_a^{(t_1)}; d_b^{(t_2)}; ...; d_e^{(t_l)}$. The combination of selected employee options for all categories (moments of decision making) constitutes a complex random event $D_j = [d_a^{(t_1)}; d_b^{(t_2)}; ...; d_e^{(t_l)}]$ that is described by the random variable *state of human factor*. In this case, the set of complex events D_1, D_2, …, D_s will form a complete group of events.

The decisions and actions of the operators affect both the probabilities of initiating events and the conditional probabilities of failures due to the corresponding initiating event. With this in mind, expression (1) can be written as:

$$P(F) = \sum_i \sum_j P(F \mid IE_i, D_j) \cdot P(IE_i \mid D_j) \cdot P(D_j) \qquad (3)$$

When various damaged end states are considered and taking into account expressions (2) the following set of equations can be formed instead of equation (3):

$$P(ES_h) = \sum_i \sum_j P(ES_h \mid IE_i, D_j) \cdot P(IE_i \mid D_j) \cdot P(D_j); \; h = 1, 2, 3, \cdots m \qquad (4)$$

Level 3. Organizational and management factors

The organizational principles and safety management procedures adopted by the company (such as the principles of conducting technical inspections, personnel selection and training procedures, principles of budgeting) are also considered random variables and are differentiated into categories $m^{(1)}$, $m^{(2)}$, …, $m^{(p)}$. The current value (implementation) of the complex event state of the organizational and management factor M_k is determined by the combination of values accepted within each category $m^{(1)}$, $m^{(2)}$, …, $m^{(p)}$ differential organizational and management factors. The set of complex events M_1, M_2,…, M_u form a complete group of events.

The values taken by organizational factors within each category are random variables that have a probabilistic influence on the variables *state of the human factor* D_j.

Therefore, the expression (3) and (4) can be written in the form (5) and (6):

$$P(F) = \sum_i \sum_j \sum_k P(F \mid IE_i, D_j, M_k]) \cdot P(IE_i \mid D_j, M_k) \cdot P[(D_j \mid M_k) \cdot P(M_k) \qquad (5)$$

$$P(ES_h) = \sum_i \sum_j \sum_k P(ES_h \mid IE_i, D_j, M_k) \cdot P(IE_i \mid D_j, M_k) \cdot P(D_j \mid M_k) \cdot P(M_k) \quad (6)$$

Level 4. Factors of safety regulation

Factors of safety regulation are also grouped into categories $g^{(1)}$, $g^{(2)}$, ..., $g^{(p)}$ etc. (such as applicable standards and norms, levels of tolerable and acceptable risk, options for economic incentives for the companies to introduce safe technologies). Next, a complex random event *factor of safety regulation* G_q is introduced, which is determined by a combination of differential factors from each category. The set of complex events G_1, G_2,..., G_e form a complete group of events.

The values of the organizational factors M_k that are in effect in the company are in a probabilistic dependence on the factors of safety regulation G_q chosen by the regulatory agency.

The nature of this dependence is determined by the probabilities $P(M_k|G_q)$ of choosing appropriate solutions in the framework of company management M_k under various options of safety regulation G_q. Thus, taking into account expressions (5) and (6), probabilistic relationships are established between the factors of safety regulation G_q and failures of the technical system.

Consequently, the probabilities of the system failure can be estimated as:

$$P(F \mid G_q) = \sum_i \sum_j \sum_k P(F \mid IE_i, D_j, M_k) \cdot P(IE_i \mid D_j, M_k) \cdot P(D_j \mid M_k) \cdot P(M_k \mid G_q) \quad (7)$$

Or when multiple end states are considered:

$$P(ES_h \mid G_q) = \sum_i \sum_j \sum_k P(ES_k \mid IE_i, D_j, M_k) \cdot P(IE_i \mid D_j, M_k) \cdot P(D_j \mid M_k) \cdot P(M_k \mid G_q) \text{ , } h = 1,2,3,\cdots m \quad (8)$$

Expressions (2), (4), (6) and (8) enable assessing the probability of the occurrence of damaged end states of the system corresponding to various options of the actions of operators and to various options of organization and management within the company and government regulatory regimes. Using special probability conversion algorithms, the quantities $P(ES_h|IE_i,D_j,M_k)$, $P(IE_i|,D_j,M_k)$, $P(D_j|M_k)$ and $P(M_k|G_q)$ and can be calculated as functions of the probabilities of intermediate events that follow the initiating event and lead to the end state. A numerical analysis of the probabilistic relationships between random variables of the multilevel model can be carried out using the mathematical apparatus of Bayesian networks and influence diagrams.

3. Application of Bayesian nets for development of multilevel models

Bayesian nets are an effective tool for the development of multi-level risk assessment models [5-7]. The following Bayesian network was constructed to assess the risk of a train derailment (figure 2). Analysis of the model should start from the bottom. At Level 1, the scenario of a technical system failure is considered: The initiating event is a brake failure, after which (depending on the actions of the dispatcher) a traffic signal for a prohibiting traffic signal may follow. Another possible factor in the accident may be the excess of the permissible speed of the train that, in combination with the incorrect operation of the turnout switch, can also create conditions for the derailment of the train. Further, the accident can develop according to various scenarios depending on whether a collision with another train or a fixed structure occurs, or whether a fire occurs or not.

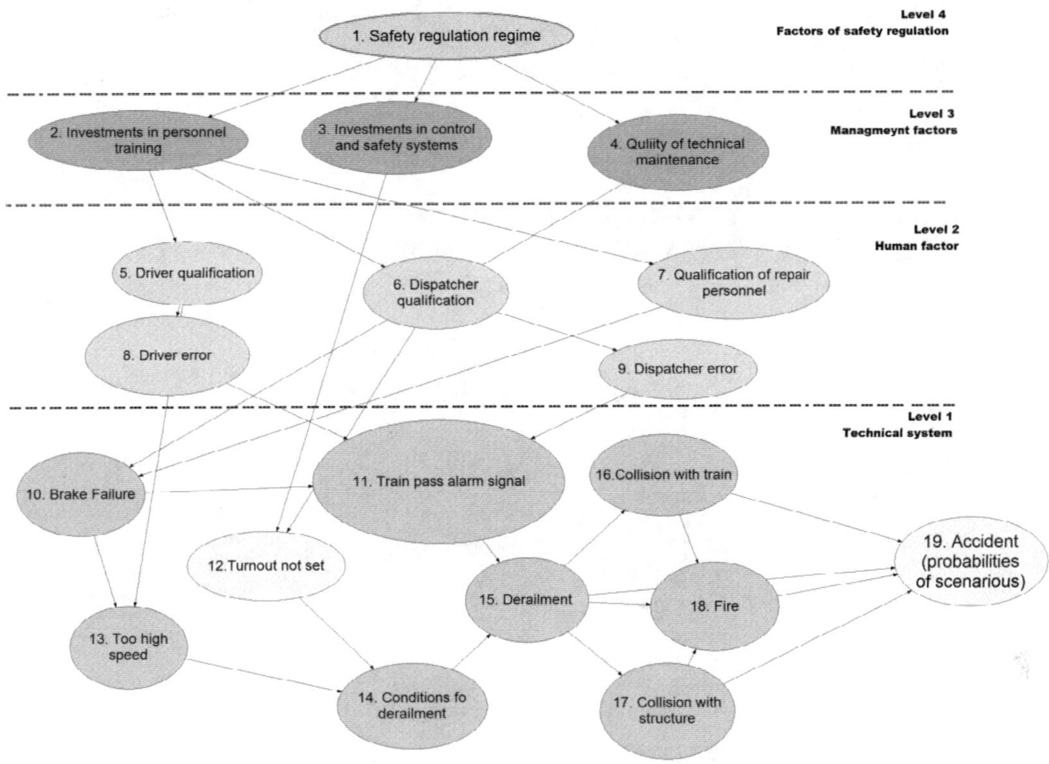

Figure 2. Bayesian network for assessment of risk of train derailment.

Level 2 presents the influence of the human factor on the development of the failure scenario. The probability of events, the scenario graph of the technical system is determined, including the variables of the state of the human factor by the quality of the maintenance procedures. The derailment of the train and subsequent events of the escalation of the accident are in probabilistic connection with the level of professional qualifications of the driver and dispatcher.

In turn, the state of the variables related to Level 2 model depends on the organization and safety management factors at the company level (Level 3). The qualifications of the personnel and the level of maintenance depend on the principles of personnel selection and training adopted by the railway company, as well as on the costs of ensuring safety.

Finally, the variables located at Level 3 are probabilistically dependent on the type of safety regulation regime at the national/sectoral level (Level 4). In this example, two regimes *liberal* and *administrative* are distinguished.

Node 1 *Safety Regulation Regime* is a decision node. The values of the variable that determines its state are selected by the regulator out of two alternatives: (1) *liberal regime* that corresponds to soft requirements and limited expenditures on ensuring safety, and (2) *administrative regime* corresponding to strict requirements and significant investments in safety systems. Node 1 *Safety regulation regime* is the root node of the diagram.

Nodes 2-19 are random network nodes. The states of each of the non-root nodes are set by the table of conditional probabilities under various states of their parents.

The probability field for the liberal regime of transportation safety regulation is presented in figure 3. The random variable *Accident* that determines the end states of the system is presented by random node 19. The probabilities of the occurrence of various accident scenarios are determined using the Bayesian net. The estimates of losses occurring in case of occurrence of various scenarios are determined on the basis of expert assessments and available statistics. The values are set based on how experts evaluate losses that correspond to various accident scenarios.

The presented diagram is part of the graphical interface of the GeNie 2.0 software package [8].

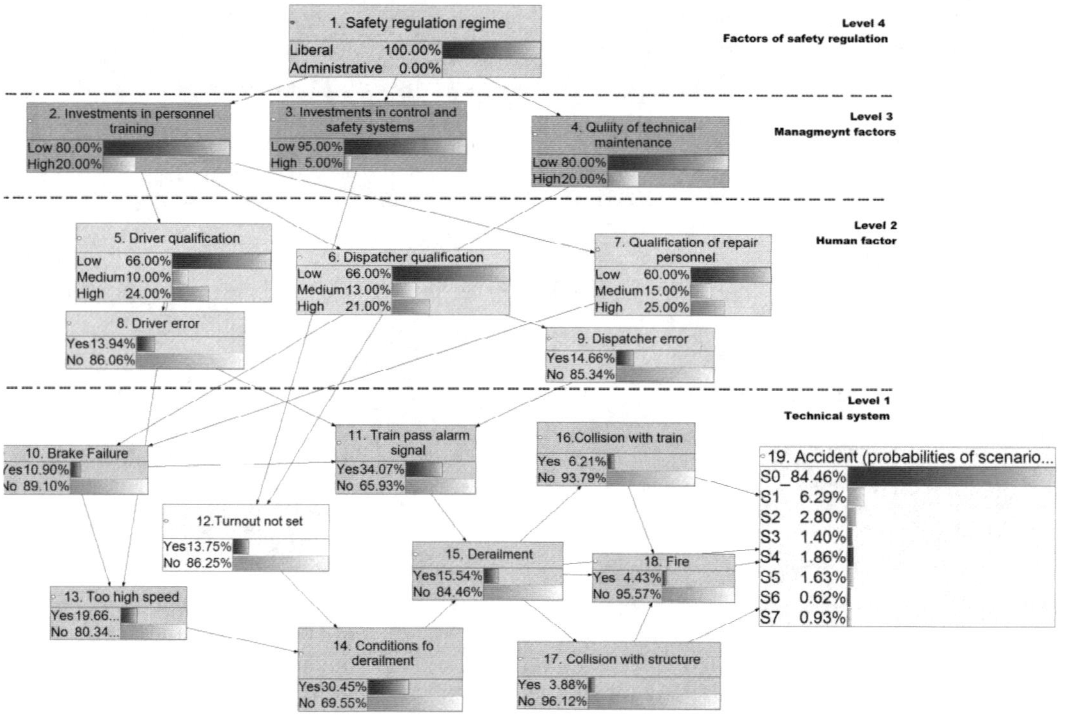

Figure 3. Probability field under the liberal regime of transportation safety regulation.

The values of risk for liberal and administrative transportation safety regulation regimes are estimated using the expression $R = \Sigma P_i U_i$ and amount to 122.61 thousand dollars under liberal and administrative regulatory regimes and 14.48 thousand dollars, respectively. The presented calculations show that in the transition from a *liberal* to an *administrative transportation safety regulation regime*, the likelihood of severe scenarios and the magnitude of risk decrease significantly. This means that by changing the factors of transportation safety regulation, a significant effect can be achieved without investments in changing the technical system.

The advantages of influence diagrams include the fact that they allow recalculating the probability field and the magnitude of risk upon additional information is received from monitoring systems about the state of individual model variables, thereby clarifying the obtained estimates. Suppose that for the example in question (figure 4) there is additional information about that under the liberal transportation safety regulation regime a *turnout error* occurred (that is, a random variable becomes deterministic), in this case the probabilistic field of the model changes significantly. In particular, an a priori estimate of the probability of a successful passage of a section decreases from 0.846 (a priori estimate) to a value of 0.603 (a posteriori estimate). At the same time, the probability of accident scenarios increase accordingly, and the posterior assessment of the risk increases to 307.90 thousand dollars compared to a priori estimate of 122.61 thousand dollars.

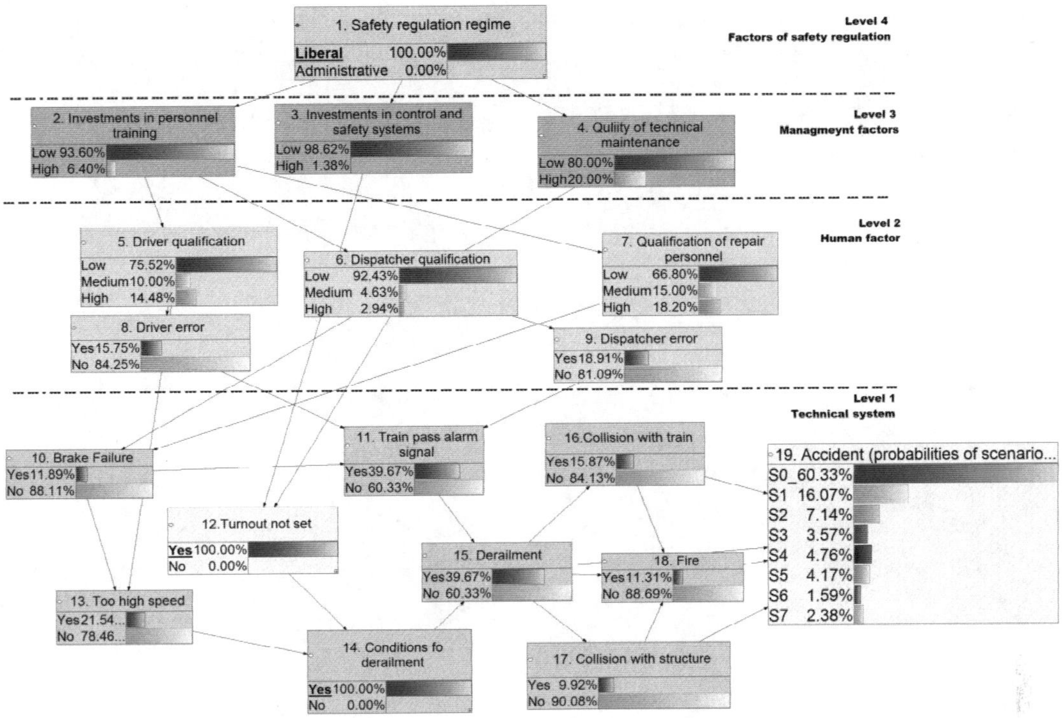

Figure 4. Updating the estimates upon receipt of additional information.

When solving risk management problems associated with the operation of a hazardous facility, it is necessary to use an integrated approach that involves a combination of actions aimed both at improving the elements of the technical system, building a more effective safety management system, and increasing the professional training of operators and personnel. To achieve it, multilevel risk assessment models must be built that allow accounting for the influence of human and organizational factors, as well as factors of safety regulation on the occurrence of various failure scenarios in the technical system.

Bayesian networks are an effective tool to analyze such multilevel models with a high level of uncertainty, as well as to refine the estimates obtained upon receipt of additional information on the values of random variables appearing in the model.

References

[1] Bea R 2002 Human and Organizational Factors in Reliability Assessment and Management of Offshore Structures. *Risk Analysis* **22** 1 29-45

[2] Pate-Cornell E and Murthy D 1996 Human and management factors in probabilistic risk analysis: the SAM approach and observations from recent application *Reliability Engineering and System Safety* **53** 2 115-126

[3] Pate-Cornell E 2002 Finding and Fixing Systems Weaknesses: Probabilistic Methods and Applications of Engineering Risk Analysis *Risk Analysis* **22** 2 319-334

[4] Makhutov N A and Reznikov D O 2019 Assessment of large-scale catastrophes in complex engineering systems *IOP Conference Series: Materials Science and Engineering* **481** 012002

[5] Charniak E 1991 Bayesian Networks without Tears *AI Magazine* **12** 50-63

[6] Pearl J 1988 Probabilistic Reasoning in Intelligent Systems: Networks of Plausible Inference Morgan Kaufmann, San Mateo, California 19882

[7] Makhutov N A and Reznikov D O 2007 The use of Bayesian networks for assessing terrorist risks and choosing the optimal strategy for countering the terrorist threat *Problems of Safety and Emergency Situations* **5** 43-63
[8] GeNle 2.0. Decision System Laboratory, University of Pittsburgh Available at: http://genie.sis.pitt.edu

Local damage to structural materials under low-cycle loading

A N Romanov

Mechanical Engineering Research Institute of the Russian Academy of Sciences, Russia

alrom37@mail.ru

Abstract. The damage accumulation kinetics in plastic (metals) and brittle materials (concrete) at the stages of fracture formation and development under cyclic loading is considered. The regularities of damage accumulation at microlevels due to structural and strain heterogeneity with the achievement of the limit state – the formation of micro-cracks are shownThe early formation curves of micro- and macro-fractures on the fatigue curves, the areas of dispersed fracture formation and development are determined.

1. Introduction

The stage of fracture formation is preceded by distributed crack formation associated with the heterogeneity of strain processes in separate volumes of the deformable material. The objectives of this study are the investigation of heterogeneity of damage formation and development in local areas of structural materials under cyclic strain with an increasing number of loading cycles and development of methods for estimating limit states at the macro- and micro levels under cyclic loading of plastic materials.

2. Material and methods

The authors used tubular samples made of single-phase austenitic steel Kh18N10T, two-phase steel 45 with a partial perlite structure, and pearlite steel TC to study the features of the local strain development and assess structural heterogeneity. Micro-hardness measurement prints were applied to the working part of the sample using a diamond pyramid on the PMT-3 device. The strain was performed under cyclic tensile-compression loading with a frequency of 1 cycle per minute. The strain on the working base was measured using a longitudinal strain-meter with an error of no more than 2% of the measured value.

Heterogeneity of cyclic plastic strain $K_{\mu\delta}$and one-way accumulated plastic strain $K_{\mu\varepsilon}$ were defined as:

$$K_{\mu\delta} = \delta_i/\delta_{av} \text{ and } K_{\mu\varepsilon} = \varepsilon_i/\varepsilon_{av} \quad (1)$$

where δ_i is the local plastic strain in the cycle; $\delta_{av} = \delta_k$ – the average value of the width of the hysteresis loop in the cycle, measured over the entire working base; ε_i – local one-way accumulated strain in the load cycle under consideration; $\varepsilon_{av} = \varepsilon_k$ – average one-way accumulated strain.

The structural heterogeneity of the material causing strain heterogeneity; can be estimated by measuring the micro-hardness of the material in the initial state.

The structural heterogeneity of the material causing strain heterogeneity; can be estimated by measuring the micro-hardness of the material in the initial state.

The heterogeneity of the structure and local strain in single-phase (steel Kh18N10T) and two-phase materials (ferrite-pearlite steel 45 and pearlite steel TS) was studied.

Based on statistical processing data, the estimation of the structural heterogeneity ratios based on the probability curves of the micro-hardness distribution is carried out. In this case, the heterogeneity ratios of the structure by scattering of micro-hardness values H_μ were determined as:

$$K_{H_\mu} = \bar{H}_\mu / H_{\mu i} \tag{2}$$

where $H_{\mu I}$ and $\bar{H}_\mu$ are the minimum and average micro-hardness values.

Following the scattering of measurement values, there is also a spread of indicators by the heterogeneity ratios of micro-hardness values (figure 1).

Heterogeneity of cyclic plastic strain $K_{\mu\varepsilon_p}$ and one-way accumulated plastic strain $K_{\mu\varepsilon}$ defined as:

$$K_{\mu\varepsilon_p} = \varepsilon_{pk}/\varepsilon_{pm} \text{ and } K_{\mu\varepsilon} = \varepsilon_k/\varepsilon_m \tag{3}$$

where ε_k is the local plastic strain in the cycle; ε_{pm} – the average value of the hysteresis loop width in the cycle measured over the entire working base; ε_k – local one-way accumulated strain in the considered loading cycle; ε_m – the average one-way accumulated strain for this cycle on a large measurement base.

The heterogeneity of the strain development, both recoverable and non-recoverable (one-way accumulating) depends on the nature of the material and its structural state, determined by the final processing.

In multiphase steels or alloys, the heterogeneity of micro-strains will be localized in the least durable stage, and damage accumulation will occur mainly in these areas, and for them, the strain heterogeneity ratios should be estimated.

3. Results and discussion

Following the scattering of measurement values, there is also a spread of indicators by the heterogeneity ratios of micro-hardness values (figure 1).

As can be seen from figure 1, a, during the first (initial) elastic-plastic strain in the half-cycle of tension (K = 0) in separate areas of 200 μm in size, with an average strain on the base of 40 mm, strains in the range from 0 to 5.0% are observed.

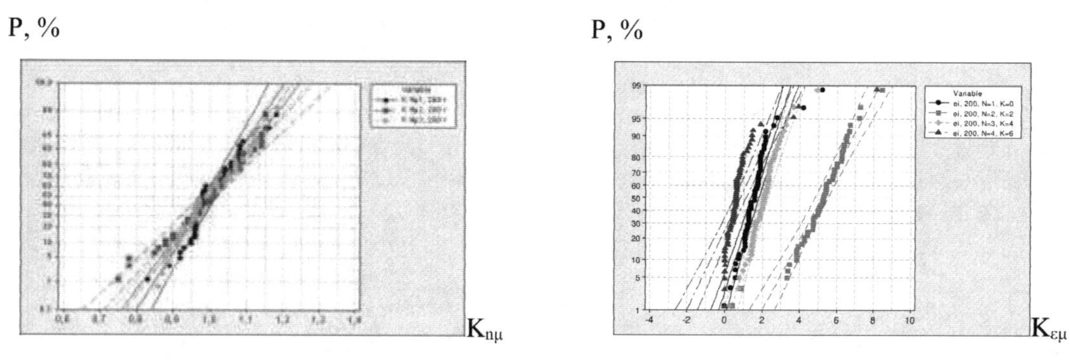

Figure 1. Distribution of micro-hardness heterogeneity ratios of TC, steel 45 and Kh18N10T steels (a) and local strains in TC steel in the first and second cycles (b).

The strain heterogeneity of the main volume (up to 80% of the accumulated probability) of two-phase steel with sufficient accuracy for practice can be considered to obey the normal distribution law. However, unlike a single-phase material, steel containing a pearlite component is deformed more heterogeneously, especially two-phase steel having both pearlite and ferritic structure (steel 45).

At 150°C exceeds the level of uniform strain at room temperature from half of the total strain of 0.5 to 0.55).

For pearlite steel TC (100% perlite), the strain in local areas is more uniform than steel with a partial share of perlite (steel 45).

The distribution of local strain and corresponding concentration ratios and values of micro-hardness in the initial state obey the normal law.

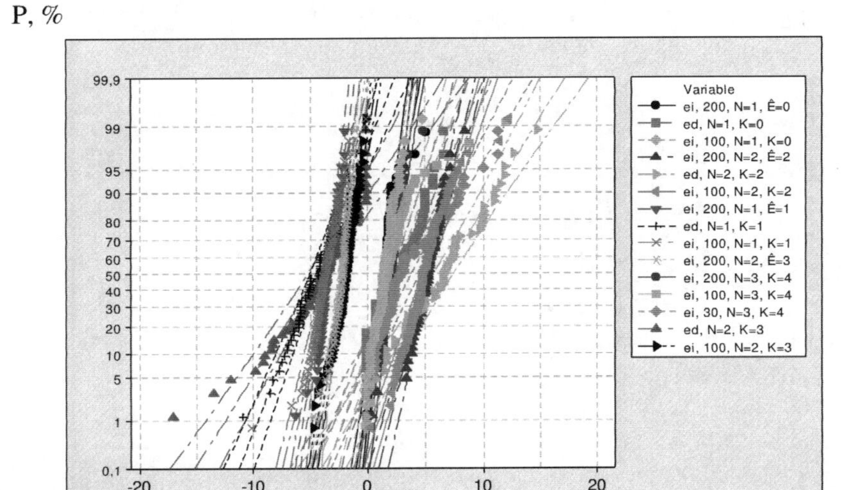

Figure 2. The statistical distribution of strains on the bases of 30 μm, 100 μm and 200 μm of TC steel in the half-cycles of expansion and contraction of the first and second cycles and the tensile half-cycle of the third cycle.

Statistical processing has established that the heterogeneity of plastic strain can be estimated using the parameters of the probability curves of the normal distribution, defined as:

$$K_{\mu\varepsilon_p} = (U_p S + \varepsilon_k)/\varepsilon_k \tag{4}$$

where U_p – quantile, S – mean square deviation, ε_{pk} – the value of the local strain in the half-cycle k, ε_k – limit strain under static fracture obtained on a large base (in this study, it was 40 mm).

For single-phase steel Kh18N10T $U_p = 2.79$, and then through the parameters of the distribution of local strain, the heterogeneity ratios can be obtained as:

$$K_{\mu\varepsilon} = \frac{(2.79S + \varepsilon_{pk})}{\varepsilon_{pk}} \tag{5}$$

The heterogeneity ratios for cyclic recoverable, one-way accumulated strain and microhardness values were the same:

$$K_{\mu\varepsilon} = K_{\mu\varepsilon_p} = K_{H\mu} = \overline{H}_\mu /(H_\mu - U_p S) \tag{6}$$

The greatest heterogeneity is observed in the first loading cycles. Further loading is accompanied by a decrease in the heterogeneity of both recoverable and accumulating strain (figure 2). The maximum strains are preserved in the same sections. The strain heterogeneity ratios are stabilized, remaining equal for both recoverable and accumulated strains.

In areas with high local strain with an increasing number of loading cycles, micro-fractures and formation of the main fracture in the surface of the material is covered with numerous micro-fractures (spread fracture formation). In single-phase materials, micro-fractures cover the loaded surfaces relatively evenly (figure 3).

For pearlite steel TC (100% perlite), the strain in local areas is more uniform than steel with a partial share of perlite (steel 45).

In two-phase materials such as steel 45, micro-fractures are formed mainly in ferritic areas (figure 4). During processing, for example, rolling, ferritic and pearlite areas take the form of strips, and micro-fractures, located in these areas, have an extended form, as shown in figure 4.

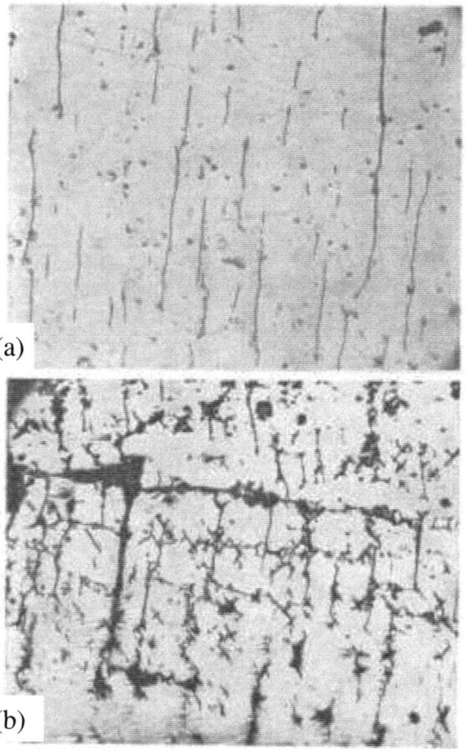

Figure 3. The nature of the formation of micro-fractures under low-cycle loading: under tension-compression (a), under torsion (b).

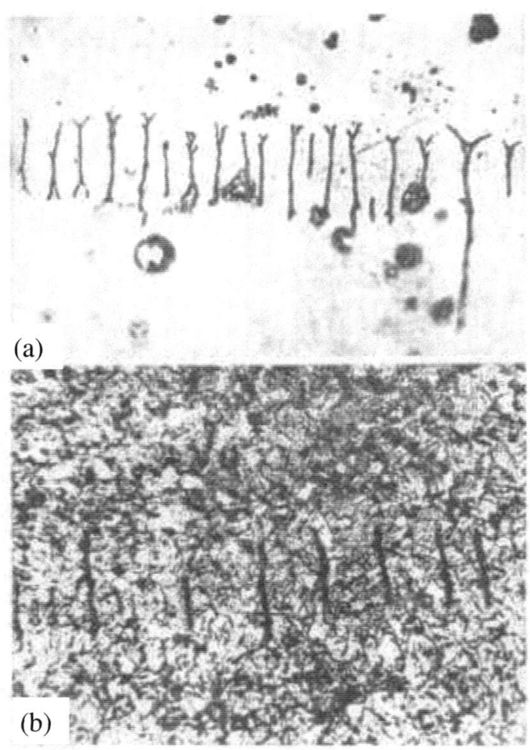

Figure 4. Micro-fracture formation of low-cycle steel fatigue in the ferritic area during cyclic torsion.

When micro-fractures reach specific sizes and densities, the main fracture is formed by fusion and interaction (figure 5), the development of which in the damaged material proceeds at an accelerated rate.

The damage accumulation in structural steels under cyclic loading is defined as [1]:

$$\eta = \int_0^N \frac{\varepsilon_p \varepsilon_{ep}}{\varepsilon_{st}^2} dN + \int_0^N \frac{\Delta\varepsilon}{\varepsilon_{st}} dN \qquad (7)$$

In the extreme case (formation of a macro-crack or loss of stability of plastic strain), the damage is described as:

$$\int_0^{N_f} \frac{\varepsilon_p \varepsilon_{ep}}{\varepsilon_{st}^2} dN + \int_0^{N_f} \frac{\Delta\varepsilon}{\varepsilon_{st}} dN = 1 \quad (8)$$

where ε_p – plastic strain in a cycle, ε_{ep}– elastic-plastic strain in the tensile half cycle, ε_{st}– strain with a single (static) fracture, corresponding to the true tensile strength, $\Delta\varepsilon$ – one-way accumulated plastic strain in the cycle, N – number of load cycles, N_f – number offracture cycles.

The evolution of the structural state and the damage accumulation kinetics at the stages of fracture formation and development in brittle materials (concrete) under cyclic loading is presented in [4,5]. Other approaches to the description of damage accumulation kinetics are considered in [4-9].

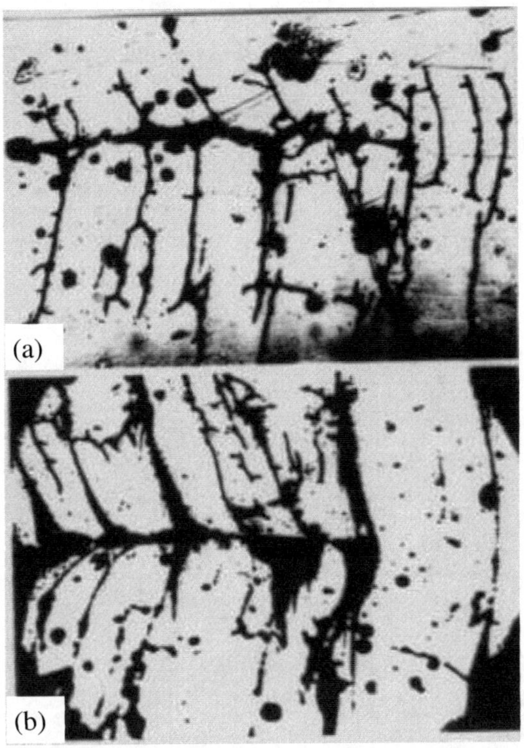

Figure 5. Formation of a main fracture in the strain heterogeneity areas under cyclic torsion.

At the stage of fracture propagation, the damage accumulation rate is estimated following the dependence [1]:

$$\eta = \int_0^N \frac{\nu_{ep} \ \nu_p}{\nu_c^2} dN + \int_0^N \frac{\Delta\nu}{\nu_c} dN \quad (9)$$

In the extreme case (fracture):

$$\int_0^{Nff} \frac{\nu_{ep}\nu_p}{\nu_{st}^2} dN + \int_0^{Nf} \frac{\Delta\nu}{\nu_{st}} dN = 1 \quad (10)$$

where ν_p– plastic (residual) exposure (margin displacement) of a fracture, ν_{ep}– elastic-plastic fracture exposure in the tensile half-cycle, $\Delta\nu_p$– the accumulated one-way exposure of fracture margins in the cycle, v_c – limit fracture exposure (margin displacement), corresponding to the maximum load in a single rupture of the sample (part) with a fracture, N and N_f – the current and destructive number of cycles, respectively.

Figure 6 shows the limit values of damage in ferrite-pearlite steel at the fracture formation stage (a) and aluminum alloy[3] at the fracture propagation stage (b). In figure 6 (b) the load in the aluminum alloy sample was reduced after the first hardening cycles (shown by triangles on figure 6 (b)).

Taking into account the heterogeneity of plastic strain under dependencies (7) and (8), the fracture condition (appearance of micro-cracks) without taking into account damage from elastic strain (quasi-static fracture) can be written as follows [2]:

$$\int_0^{N_f} \frac{(K_\mu \varepsilon_p)^2}{\varepsilon_{st}^2} dN + \int_0^{N_f} \frac{K_{\mu\varepsilon} \Delta\varepsilon_p}{\varepsilon_{st}} dN = 1 \quad . \tag{11}$$

Figure 7 shows the fatigue curve of ferritic-pearlite steel under loading with a given range of elastic-plastic strain (hard loading). The red line represents the moments of formation of initial micro-cracks, determined by the dependence (6) with the strain heterogeneity ratio equal to 1.1. Dash blue lineindicates the moments of macro-crack formation depending on the number of fracture cycles. The areaindicates the moments of macro-crack formation depending on the number of fracture cycles.

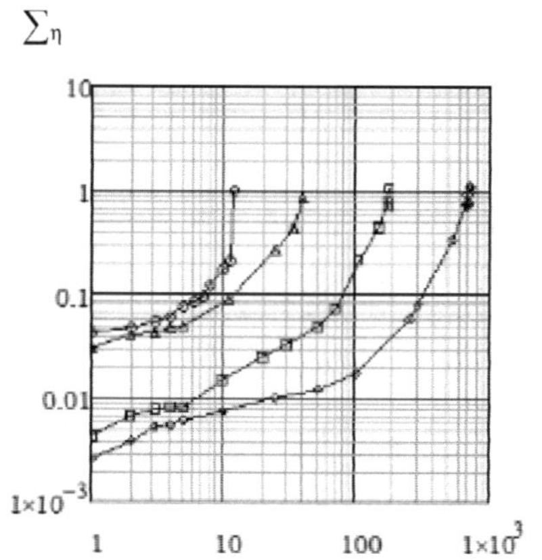

(a)

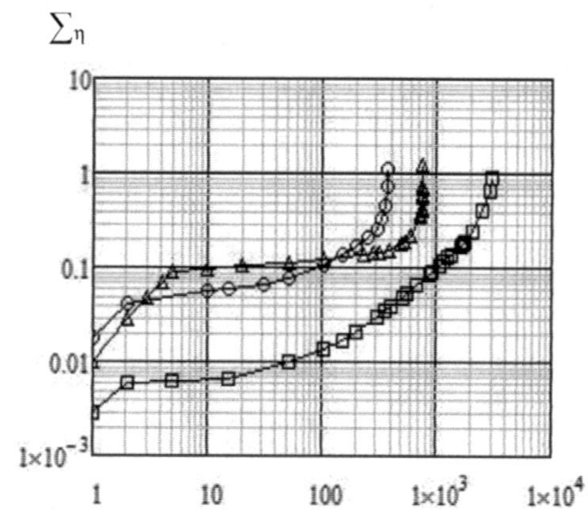

(b)

Figure 6. Limit damage accumulations calculated by criterion (8) of ferrite-pearlite steel (a) and calculated by criterion (10) of aluminum alloy AD33 (b).

The area between the red and the dashed line is the area of dispersed crack formation, the ultimate development of which is the formation of a macro-crack. The area between the dashed blue line and the fatigue curve (black line) is the area where the main macro-crack propagates until complete fracture.

Thus, the studies have shown that the nature of the development of strain processes and the nature of damage accumulation is determined by the structural state of the deformable material and its phase composition.

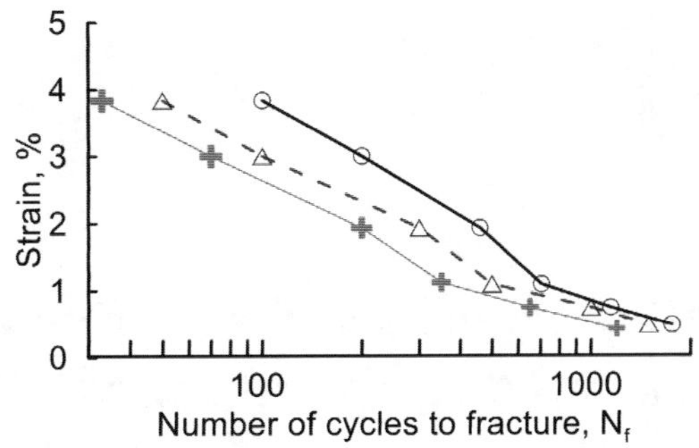

Figure 7. The fatigue curve of ferrite-pearlite steel under rigid loading.

4. Conclusion

1. The strain heterogeneity of materials depends on their structural heterogeneity and can be estimated from the indicators of structural heterogeneity determined by micro-hardness measurements.

2.Damage accumulation kinetics and limit states at the micro-level (micro-crack formation) and macro-level (main macro-crack formation) under cyclic loading of plastic materials (metals).

3.The ability to determine the limit states at the micro- and macro-levels allows indicating the limit state curves on the fatigue curves (the beginning of the formation of micro-and macro-cracks).

References

[1] Romanov A N 2007 *Inzh. Zh. Sprav. Prilozh.* **6** 35
[2] Romanov A N and Filimonova N I 2020 *J. Phys.: Conf. Series* **1431** conf. 1 No 012049
[3] Romanov A N, Nesterenko G I and Filimonova N I 2018 *J. Mach. Manuf. Reliab.* **47**(5) 414–19
[4] Fu X, Zhang J and Lin J 2018 *Fatigue Fract. Eng. Mater. Struct.* **41**(5) 1077
[5] Du W, Luo Y, Wang Y, Chen S and Yu D 2017 *Fatigue Fract. Eng. Mater. Struct.* **40**(6) 994
[6] Xia T, Zou J, Gao D and Yao W 2016 *Fatigue Fract. Eng. Mater. Struct.* **39**(2) 194
[7] Shah M, Rezai-Aria Fand Mabru C 2015 *Fatigue Fract. Eng. Mater. Struct.* **38**(6)742
[8] Socha G and Dietrich L 2014 *Fatigue Fract. Eng. Mater. Struct.* **37**(2) 195
[9] Tankasala H C, Deshpande V S and Fleck N A 2018 *Int. J. Fracture* **21**(2) 123

Acknowledgments

The work is executed at the support of the Russian Foundation for basic research (project No. 16-58-48008 Ind_ome).

Damage accumulation in cyclically stable steel under low-cycle loading and elevated temperature

A N Romanov

Mechanical Engineering Research Institute of the Russian Academy of Sciences, Russia

alrom37@mail.ru

Abstract. Based on the strain-kinetic fatigue fracture criterion, the kinetics of accumulated damages in structural carbon steel under low-cycle loading at an elevated temperature (150°C), at which the material begins to develop stress aging processes, is estimated. It is shown that stress aging, strengthening the material, narrows the durability area in which there is quasistatic destruction, the limiting case of which is the buckling failure of plastic strain, like single static destruction. This increases the fatigue fracture range, the limiting case of which is the formation of the primary fracture. The recoverable plastic strain causes the main damaging effect. Stress aging contributes to an increase in damage from elastic strain, although in the area of low-cycle failure, the share of damage from elastic strain remains small, growing with increased durability, and in the area of multi-cycle fatigue, this share becomes prevalent. Hardening from stress aging inhibits the development of damage from the one-way accumulated strain. The author confirms the validity of applying the strain-kinetic criterion of fatigue fracture to describe the damage kinetics and limit states during the cyclic elastic-plastic strain of carbon steel in the presence of weak stress aging.

1. Introduction

Damage accumulation kinetics depends on the structural state of the structural material and strain conditions.

As is known [1], structural materials can be strengthened (hysteresis loop width decreases), softened (hysteresis loop width increases), or stabilized (hysteresis loop width remains almost constant) with an increase in the number of elastic-plastic strain cycles.

Structural changes, such as stress aging, can significantly change the plastic properties of a material under load and thereby affect the damage accumulation kinetics and the limit state [1]. Structurally unstable materials can also increase their plastic properties due to changes in the structural state under cyclic loading, and then changes in the plasticity of the material must be taken into account when determining the damage accumulation kinetics and evaluating the limit state [2].

Under low-cycle loading, there can be two limiting states: the formation of macro-fissures or buckling failure of plastic strain. The latter proceeds by the type of static destruction, and in this regard it is called quasi-static.

Quasi-static fracture is characteristic of cyclic elastic-plastic strain with a given load amplitude or with an asymmetric loading cycle. When deformed with a given range of elastic-plastic strain, the destruction in all cases occurs with the fracture formation.

Quasi-static fracture is characteristic of a small number of loading cycles, which depends on the load level. Fatigue fracture (resulting from fracture formation) occurs during strain with a uniform strain before fracture.

The purpose of this work is to identify the damage accumulation kinetics in carbon steel and the possibility of applying the strain-kinetic criterion of fatigue fracture of carbon steel 22k for an elevated temperature at which the specified steel begins to develop stress aging processes that reduce the plastic properties of the material, increasing its strength indicators.

2. Materials and methods

Carbon steel 22k (chemical composition in %: C 0.22, Si 0.3, Mn 0.8, Ni 0.3, S 0.4, P 0.4, Cr 0.4, Ti 0.05, Cu 0.3, Fe – base; mechanical properties: σ_b = 530 MPa, $\sigma_{0.2}$ = 265 MPa, Ψ = 46%) at room temperature is cyclically stable. Cyclically stabilizing materials are characterized by the value of uniform strain (strain corresponding to the conditional tensile strength σ_b) [1]. It is widely used for manufacturing machine-building parts and assemblies.

Tests for low-cycle fatigue were carried out under tension-compression with a frequency of 3–5 cycles per minute with a symmetrical load cycle (soft loading) and with a given range of elastic-plastic strain (hard loading) at a temperature of 150°C. The sample is solid, corset type with a diameter of 12 mm. Transverse strains were measured, which were converted to longitudinal ones using the transverse strain ratios: 0.5 – for plastic strain and 0.3 – for elastic strain. The error in measuring strains did not exceed 0.5% of the measured value. The tests were carried out in a vacuum chamber ~ $5\cdot10^{-2}$ mm Hg placed on the columns of the testing machine.

Evaluation of damage accumulation levels with fractions of damage from cyclic recoverable (hysteresis loop width in the tensile half cycle), one-sided accumulation of plastic and elastic strains was carried out based on the previously proposed strain-kinetic criterion of fatigue fracture [1] without considering the damaging role of elastic strain for the current number of cycles (mainly quasi-static fracture) as:

$$\eta = \int_0^N \frac{\varepsilon_p^2}{\varepsilon_{st}^2} dN + \int_0^N \frac{\Delta\varepsilon}{\varepsilon_{st}} dN \tag{1}$$

In the extreme case (formation of a macrofissure or loss of stability of plastic strain) as:

$$\int_0^{N_f} \frac{\varepsilon_p^2}{\varepsilon_{st}^2} dN + \int_0^{N_f} \frac{\Delta\varepsilon}{\varepsilon_{st}} dN = 1 \tag{2}$$

For the entire range of fatigue fracture durations (low-cycle and multi-cycle fatigue) as:

$$\eta = \int_0^N \frac{\varepsilon_p \varepsilon_{ep}}{\varepsilon_{st}^2} dN + \int_0^N \frac{\Delta\varepsilon}{\varepsilon_{st}} dN, \tag{3}$$

Fracture as:

$$\int_0^{N_f} \frac{\varepsilon_p \varepsilon_{ep}}{\varepsilon_{st}^2} dN + \int_0^{N_f} \frac{\Delta\varepsilon}{\varepsilon_{st}} dN = 1. \tag{4}$$

With the reflection of the damaging role of elastic strain, dependencies (3) and (4) can be written as:

$$\eta = \int_0^N \frac{\varepsilon_p^2}{\varepsilon_{st}^2} dN + \int_0^N \frac{\Delta\varepsilon}{\varepsilon_{st}} dN + \int_0^N \frac{\varepsilon_p \varepsilon_e}{\varepsilon_{st}^2} dN \tag{5}$$

and when fractured:

$$\int_0^{N_f} \frac{\varepsilon_p^2}{\varepsilon_{st}^2} dN + \int_0^{N_f} \frac{\Delta\varepsilon}{\varepsilon_{st}} dN + \int_0^{N_f} \frac{\varepsilon_p \varepsilon_e}{\varepsilon_{st}^2} dN = 1 \tag{6}$$

where ε_p is the plastic strain in the tensile half cycle, ε_{ep} is the elastic-plastic strain, ε_e is elastic strain equal to $\varepsilon_e = \sigma_a/E$, ε_{st} is the static strain of a single fracture corresponding to the true tensile strength (rupture strength) to the time of buckling, plastic strain (beginning of necking), N is the current number of loading cycles, N_f is the number of cycles to fracture (formation of cracks).

Depending on (1)–(6), the first term defines fatigue damage from the action of cyclically recoverable plastic strain, the second term – damage from unilateral plastic strain, and the third term – damage from the action of elastic strain. When loading with a given range of elastic-plastic strain (hard loading), the second term in the dependencies (1–6) is zero (strain is not accumulated).

For e_{yn} = const and ε_p= const (for example, for a cyclically stable material or when changes in the hysteresis loop width and elastic strain during loading can be ignored), the dependence (4) can be written as:

$$\varepsilon_p e_{yn} N_p = \varepsilon^2 \quad \text{or} \quad \frac{\varepsilon_p e_{yn}}{\varepsilon^2} N_p = 1 \tag{7}$$

In dependencies (5), (6), (8), (9) the first term is damage from recoverable plastic strain ε_p (strain in a half cycle of tension), the second term is damage from unilaterally accumulating strain $\Delta\varepsilon$ (quasistatic damage), and the third term is damage from the action of elastic strain ε_e.

The criterion allows determining the levels of accumulated damage in each cycle and the total accumulated over the entire range of durations (both in the low-cycle and multi-cycle areas) and allows describing the limit states (formation of macro fissures or buckling failure of plastic strain). Other approaches to describing the damage accumulation kinetics are presented in [3–8].

3. Results and discussion

As follows from figure 1 (a), when the temperature increased from room to 150 °C (figure 1 (a)), the material showed some tendency to harden (uniform strain at 150°C exceeds the level of uniform strain at room temperature from half of the total strain of 0.5 to 0.55).

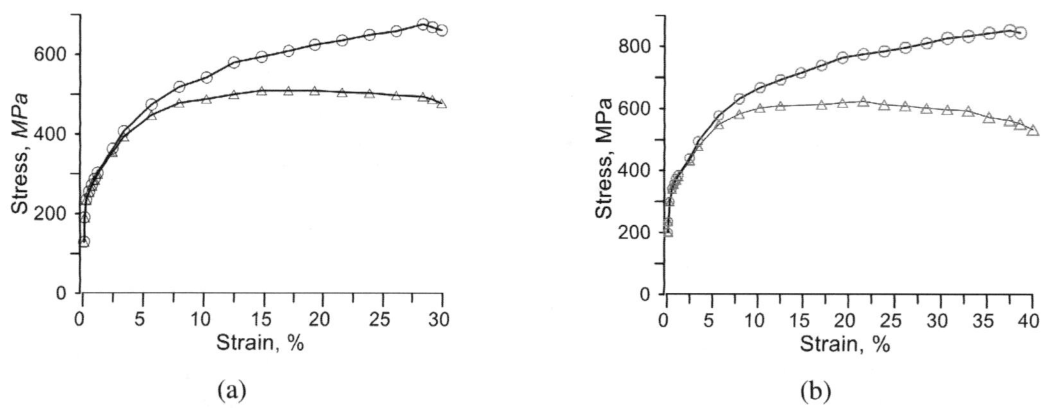

Figure 1. Static tension curves of steel 22k at 150°C (a) and at room temperature (b): curves in conditional (triangles), and in true stresses (circles).

For steel 22k at 150°C, the limit state (buckling failure of plastic strain) was observed at a strain of about 30% (figure 1 (a)). For room temperature, the buckling failure occurred at a strain equal to 37% (figure 1 (b)).

At a temperature of 150°C, the steel showed (figure 2 (a)), as at room temperature [1], significant hardening after initial loading (in the second cycle). The subsequent cyclic strain is characterized by softening (increase in strain in the half-cycle of tension), the degree of which decays to the number of cycles that make up about half of the durability of each sample. At room temperature, this stage shows the relatively uniform strain. A further increase in the number of loading cycles (loading time) begins to manifest the processes of stress aging, accompanied by a hardening of the material (recoverable strain decreases).

The processes of hardening and softening also affect the rate of accumulation of residual strains (figure 2 (b)). At this stage of hardening, there is even a slight slowdown in accumulation (strain in the compression half-cycles is greater in absolute value than in the tension half-cycles). At room temperature, accumulation at this stage practically does not occur [1]. Starting from the level of maximum recoverable strain (figure 2 (a)), the processes of unilateral strain accumulation are activated (figure 2 (b)). At the same time, the processes of stress aging are becoming more and more active (recoverable strain is continuously decreasing). This stage of loading is characterized by active strain accumulation even at room temperature [1].

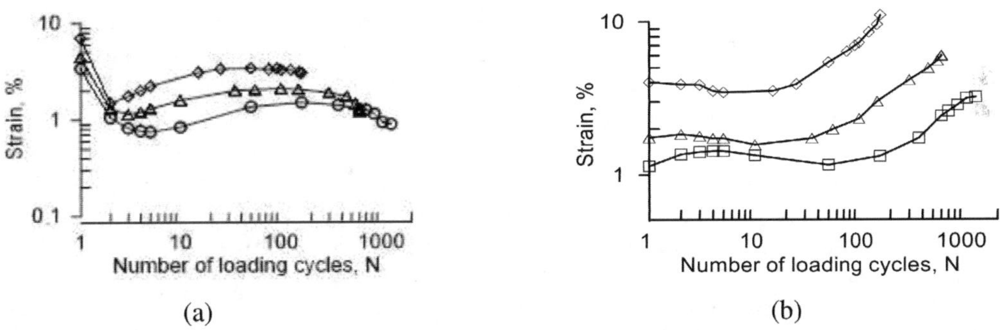

Figure 2. Strain kinetics in the half-cycle of tension (a) and accumulated strain (b) of 22k steel at a temperature of 150 °C.

Strengthening of the material from plastic strain and stress aging is accompanied by an increase in true stresses (figure 3), determined by taking into account the change (decrease) in the working cross-section of the sample from the current and accumulated strain in each cycle.

At a temperature of 150°C, significant hardening after the initial half-cycle (figure 2 (a)), and a decrease in the working cross-section, a sharp increase (jump) in true stresses is observed (figure 3). Further loading under changes in the loop width (strain in the half-cycle of tension) and one-sided accumulated strain is characterized by a slight change in the true stresses. Moreover, only in the second half of the durability, there is an active increase in true stresses.

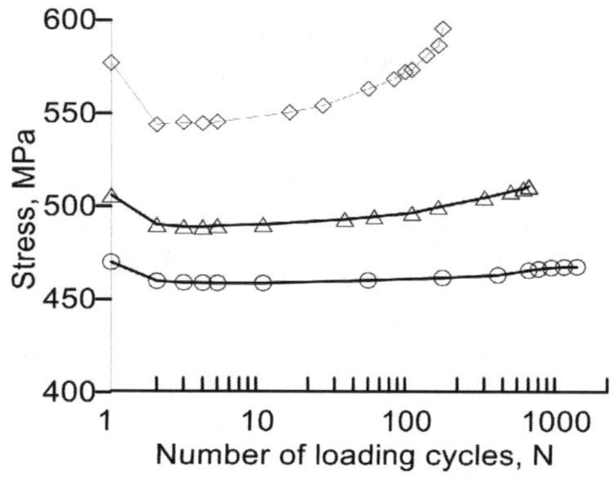

Figure 3. True stresses kinetics in 22k steel under low-cycle loading at a temperature of 150°C.

At room temperature, on the contrary, after strain in the initial half-cycle and an extensive hardening only due to plastic strain (temperature stress aging is absent or not large) and the resulting decrease in the working cross-section, hardening occurs after the initial loading [1]). Subsequent loading does not cause significant changes in the true stresses, except for quasistatic failure (N_p = 165 cycles), when large plastic strain develops a one-sided accumulation of plastic strain and a significant decrease in the working cross-section of the sample.

The nature of changes in the strain characteristics under low-cycle loading determines the damage accumulation kinetics. From figure 4, it can be seen that for durations from 160 (figure 4 (a)) to 2050 cycles (figure 4 (c)), the main damage to the moment of fracture (limit state) accumulates from the recoverable plastic strain. Although at the initial stage, the damage was maintained constant from the accumulated strain in the first cycle. As the number of cycles increases, the accumulation from recoverable strain (strain in half-cycles of tension) begins to prevail and becomes the main one by the time of fracture. The equality of damage from recoverable strain and accumulating with durability N_p= 165 cycles is equal to loading cycles (figure 4 (a)). For N_p = 610 cycles, the comparison of recoverable and accumulated damage is observed for loading cycles (figure 4 (b)). The durability of 2050 cycles shows the equality of the specified strain for loading cycles (figure 4 (b)). Damage from the elastic strain with a small number of loading cycles is not large at the initial stage, but the rate of damage growth is high. With increasing durability (N_p = 610 cycles), equality of accumulated and elastic strain is achieved by the moment of failure (figure 4 (b)). If the durability is N_p = 2050 cycles, this equality is equal to cycles and is equal to % of the number of failure cycles (figure 4 (b)).

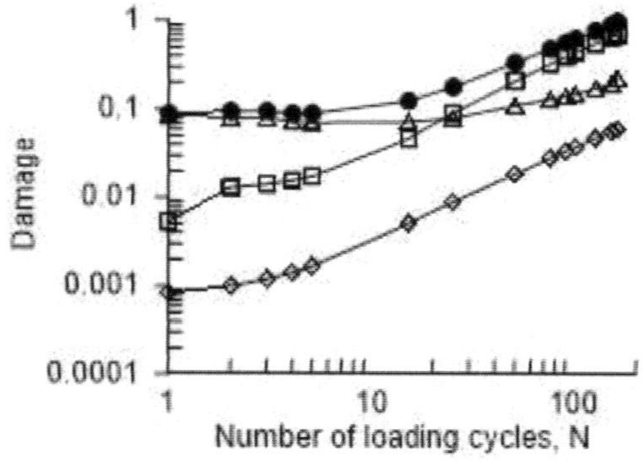

(a)

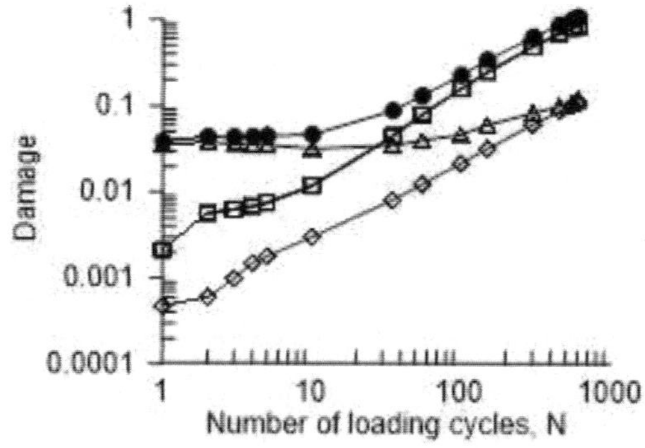

(b)

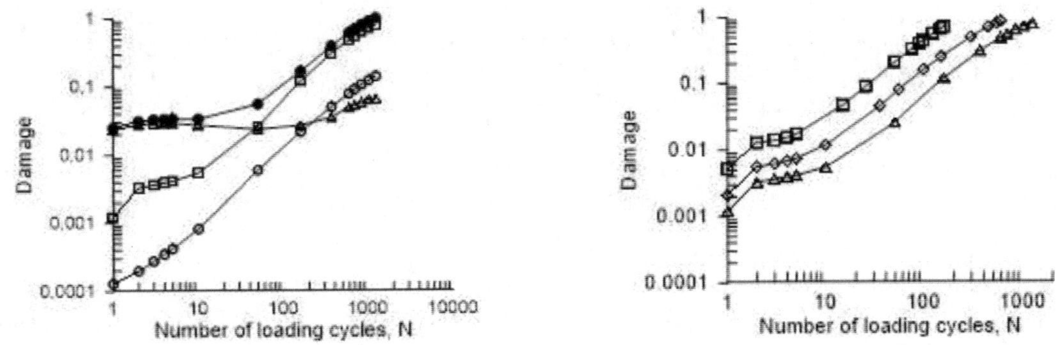

(c) (d)

Figure 4. Damage accumulation from elastic (rhombus), recoverable plastic (square), accumulating (triangle) and total (black circle) damage to each sample (a-c) and damage from recoverable strain of three samples (d): a – σ_a = 515 MPa. N_p = 165 cycles; b – σ_a = 475 MPa. N_p = 610 cycles; c – σ_a = 449 MPa. N_p == 1257 cycles.

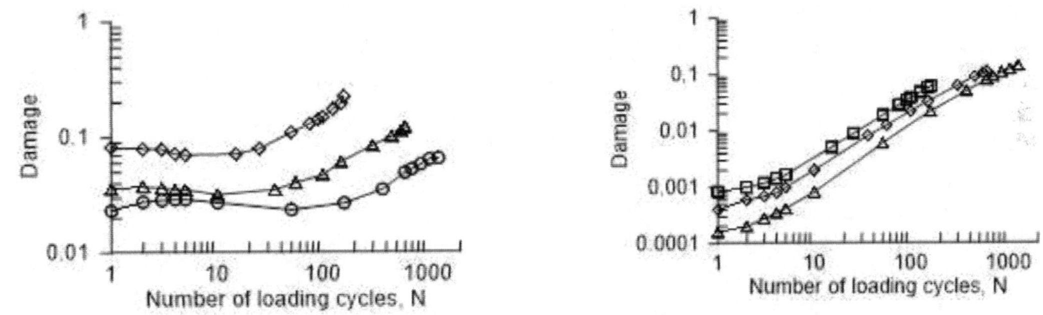

(a) (b)

Figure 5. Damage from accumulated (a) and elastic (b) strains of 22k steel at a temperature of 150°C in soft loading.

The damage accumulation kinetics from recoverable, accumulated, and elastic strain under low-cycle loading at room temperatures [1] differs significantly from accumulation at 150°C (figure 5 (a)). If the maximum accumulated damage from recoverable strain for the considered duration is in the range of 70-80% (figure 5 (a)), then damage from accumulated strain is no more than 20% (figure 5 (b)), and for elastic strain does not exceed 10% (figure 5 (b)).

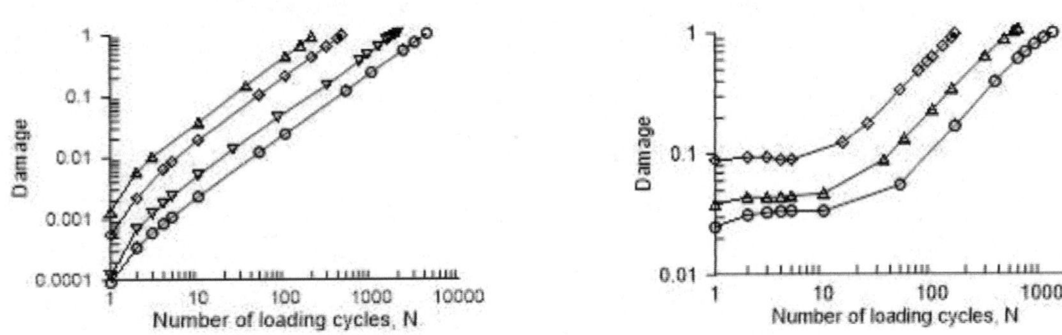

(a) (b)

Figure 6. Maximum damage accumulation in 22k steel at 150°C for hard (a) and soft loading (b).

When loading with a given range of elastic-plastic strain (hard loading), the kinetics of damage accumulation is linear (figure 6 (a)) and is well described by the strain criterion (7). For figure 6 (a), the first cycle shows damage to the original half-cycle, which caused some nonlinearity in the first cycle.

At the same time, strain with a given load amplitude shows a nonlinear nature of damage accumulation with an increase in the number of loading cycles (figure 6 (a)) and their summation obeys the dependence (6).

4. Conclusions

1. Resistance to low-cycle loading of carbon steel at elevated temperature (150°C) is characterized by an increase in strength properties due to the development of stress aging processes.

2. Stress aging slows down the rate of increase in recoverable and accumulated strains, contributing to the growth of durability.

3. Strain-kinetic criterion of fatigue fracture describes the damage accumulation kinetics and the limit state in the presence of stress aging.

References

[1] Romanov A N and Filimonova N I 2019 *J. Phys. Conf. Ser.* **1431** 012049
[2] Romanov A N and Filimonova N I 2019 *J. Mach. Manufact. Reliab.* **48** 442–47
[3] Fu X, Zhang J and Lin J 2018 *Fatigue Fract. Eng. Mater. Struct.* **41** 1077
[4] Du W, Luo Y, Wang Y, Chen S and Yu D 2017 *Fract. Eng. Mater. Struct.* **40** 994
[5] Xia T, Zou J, Gao D and Yao W 2016 *Fatigue Fract. Eng. Mater. Struct.* **39** 194
[6] Tankasala H C, Deshpande V S and Fleck NA **2018** *Int. J. Fract.* **21** 123
[7] Pervaiz Fathima K M and Chandra Kishen J M 2015 *Arch Appl Mech* **85** 921–936 DOI 10.1007/s00419-015-1001-z
[8] Pervaiz Fathima K M and Chandra Kishen J M 2013 *Int. J. Fatigue* **54** 17–24

Acknowledgments

This work is performed with financial support of the Russian Federal Property Fund (Ind_omi grant No. 16-58-48008)

Nonlinearity of the stress-strain dependence of quasi-isotropic CFRP laminate under re-static loading

P B Severov

Mechanical Engineering Research Institute of the Russian Academy of Sciences, Moscow, Russian Federation

alpinprom@yandex.ru

Abstract. As part of the task of studying the degradation of material properties under re-static loading of a quasi-isotropic CFRP laminate the nonlinearity of the stress-strain dependence in the areas of active deformation is proved. The optimal curve approximating the experimental data from the point of view of the problem is chosen. The change in the derivatives of this curve can be use σ to assess the degradation of the material properties during re-loading.

1. Introduction

Studying the behavior of CFRP laminates in the field of applied forces from the standpoint of a system evolving to a critical state [1], it is necessary to determine several measurable physical quantities, the change of which will make it possible to assess the degradation of material properties with an increase in the number of cycles of re-static loading. The task is to use the nonlinear properties of CFRP laminate itself as an indicator of degradation, if the assumed nonlinearity and its subsequent development can be established. The number of measured physical quantities is minimal – "time", "stress" and "strain".

The report focuses on the first part of the problem posed - a functional description of the experimental stress-strain dependence when testing a specimen of a quasi-isotropic structural FCRP laminate in one of the cycles of re-static loading.

2. Materials and test methods

The specimen is cut from a quasi-isotropic FCRP laminate with $[0/\pm45_2/0_2/90/0_2/\pm45_2/0]_s$ staking in the 0 direction. The dimensions of specimen working zone are as follows: length 140 mm, cross section 50x6.03 mm^2, center hole Ø14 mm.

The test was carried out in hydraulic grips of the INOVA IK-6033 servo-hydraulic testing machine. The system for collecting and processing experimental information was based on electronic equipment and software from National Instruments (USA).

Quasi-static uniaxial tension with displacement control S in accordance with figure 1. The duration of odd sections is 125 seconds, and even sections - 50 seconds. The modulus constant velocity of grip displacement on odd sections V_s = 4.4 μm·s^{-1}, on even sections V_s = 0 μm·s^{-1} (figure 1, 2). Figure 3 shows an asymmetric change in the nominal stress σ (hereinafter referred to as "stress") in the cross section of a specimen 50x6.03 mm^2 depending on time. An extensometer with a 20 mm base was installed on the side plane of the specimen symmetrically to the central hole. The average strain ε was measured on a 20mm base of the extensometer (hereinafter referred to as "strain").

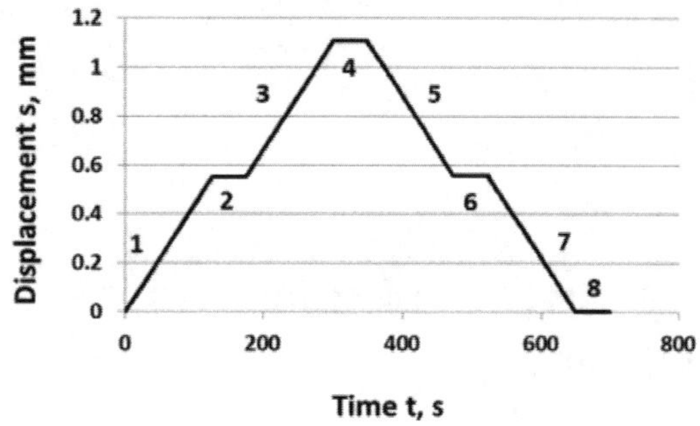

Figure 1. Specimen loading diagram.

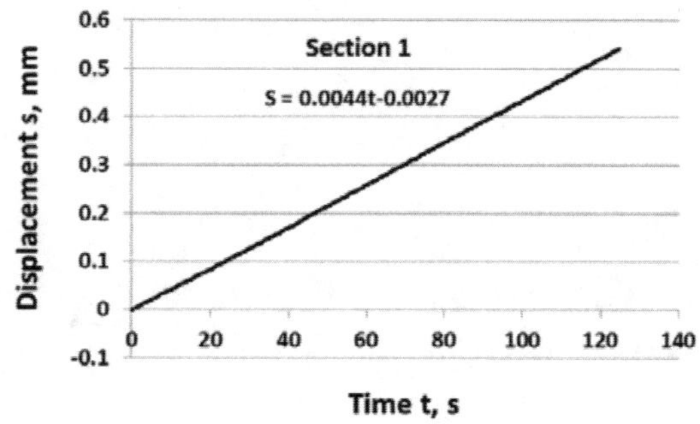

Figure 2. Section 1 of the specimen loading diagram, Vs=4.4 μm·s^{-1}.

3. Research results

The experimental points (ε_i, σ_i) under loading of the specimen in accordance with figure 1 are shown in figure 4.

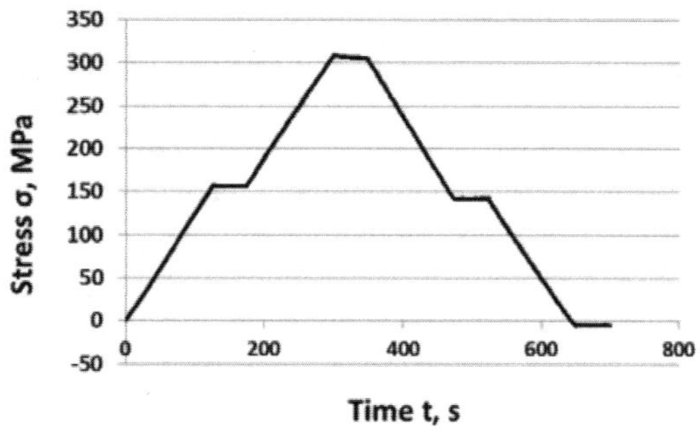

Figure 3. Asymmetric stress change in the cross section of the sample versus time.

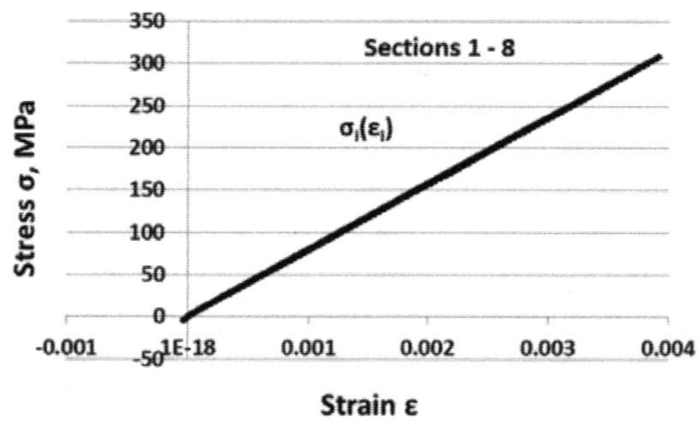

Figure 4. Specimen loading diagram (figure 1): experimental points (ε_i, σ_i).

As a result of linear approximation of the entire array of points (ε_i, σ_i) in the sections of the loading diagram from the 1-st to the 8-th inclusive, the averaged equation has been obtained:

$$\sigma = 78724 \cdot \varepsilon - 0.3074 \tag{1}$$

When considering the experimental points (ε_i, σ_i) on each separate section of the specimen loading diagram (figure 1), it is established that the dependences between them are individual and cannot be described by equation (1). As an illustration, figures 5 and 6 in an enlarged scale show the experimental points (ε_i, σ_i) at minimum and maximum strains, respectively. Sections 1, 3, 6, 8 of the loading diagram belong to active deformation (σ↑), and sections 2, 4, 5, 7 to passive deformation (σ↓) [2, 3].

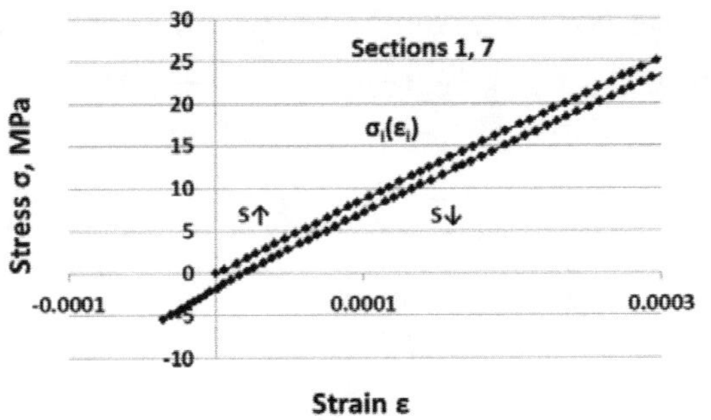

Figure 5.Specimen loading diagram (figure 1): experimental points (ε_i, σ_i) at minimum strain.

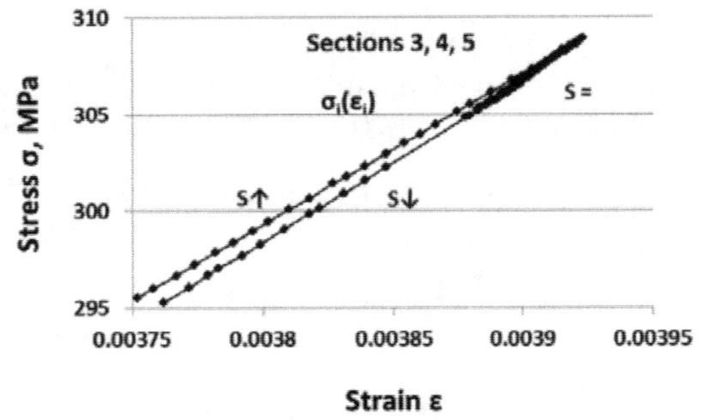

Figure 6.Specimen loading diagram (figure 1): experimental points (ε_i, σ_i) at maximum strain.

To prove the nonlinearity of the experimental dependence $\sigma_i = f(\varepsilon_i)$ and its subsequent mathematical description, we select the points (ε_i, σ_i) of the 1-st and 3-rd sections of the active ($\sigma\uparrow$) deformation of the specimen, combining them into a single array of the conditional stress-strain diagram $\sigma_i = f^*(\varepsilon_i)$ (figure 7). As a result of linear approximation of the specified array of points, we obtain an averaged equation

$$\sigma = 78334 \cdot \varepsilon + 1.7454 \quad (2)$$

in which the coefficient at ε (elastic modulus) is slightly less than the coefficient at ε in equation (1), which is evident. The nonlinearity of the experimental dependence $\sigma_i = f^*(\varepsilon_i)$ is confirmed by the behavior of the array of points (figure 8), each of which is a tangent modulus $E(\varepsilon_i)$ calculated from three adjacent points of the conditional stress-strain diagram $\sigma_i = f^*(\varepsilon_i)$, with subsequent one point advance in the direction of increasing strain.

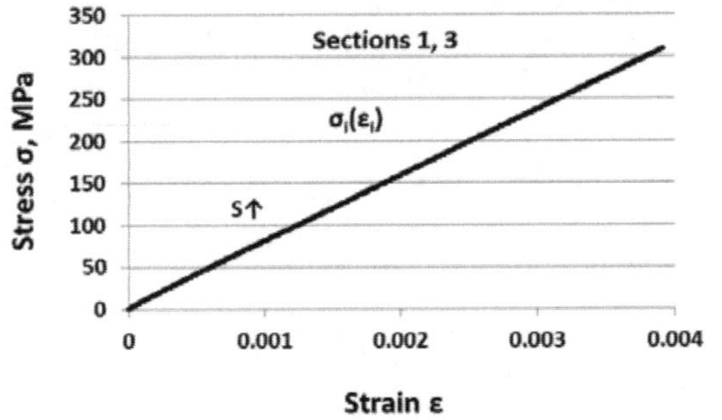

Figure 7. 1-st and 3-rd sections of the specimen loading diagram: experimental points (ε_i, σ_i).

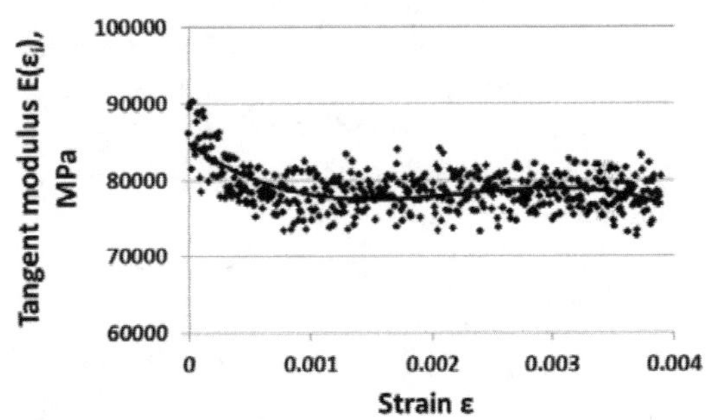

Figure 8. 1-st and 3-rd sections of the specimen loading diagram: an array of values of the tangent modulus $E(\varepsilon_i)$.

The graph of the curve

$$E(\varepsilon) = -8.484 \cdot 10^{11} \cdot \varepsilon^3 + 5.782 \cdot 10^9 \cdot \varepsilon^2 - 1.168 \cdot 10^7 \cdot \varepsilon + 8.486 \cdot 10^4 \quad (3)$$

approximating the specified array of points is shown in figure 9. The tangent modulus $E(\varepsilon)$ changes during deformation, the experimental dependence $\sigma_i = f^*(\varepsilon_i)$ is nonlinear. By simple integration of equation (3), it is possible to obtain the equation of the conditional stress-strain diagram $\sigma = f^*(\varepsilon)$ in the form of a polynomial of the 4-th degree.

The equation of the conditional stress-strain diagram $\sigma = f^*(\varepsilon)$ in the form of a polynomial of the 4-th degree can also be obtained by approximating the original array of points (ε_i, σ_i) (figure 7):

$$\sigma = -1.535\cdot 10^{11}\cdot\varepsilon^4 + 1.358\cdot 10^9\cdot\varepsilon^3 - 4.046\cdot 10^6\cdot\varepsilon^2 + 8.279\cdot 10^4\cdot\varepsilon + 5.360\cdot 10^{-1} \quad (4)$$

In this case, the slope of the function graph $\sigma = f^*(\varepsilon)$ to the positive direction of x-axis (figure 10) is determined by the first derivative of the function (4) at the strain ε:

$$d\sigma/d\varepsilon = -6.140\cdot 10^{11}\cdot\varepsilon^3 + 4.074\cdot 10^9\cdot\varepsilon^2 - 8.092\cdot 10^6\cdot\varepsilon + 8.279\cdot 10^4 \quad (5)$$

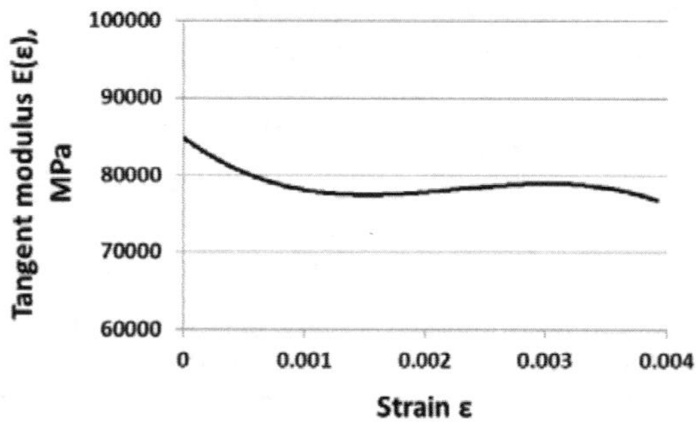

Figure 9.1-st and 3-rd sections of the specimen loading diagram: approximating curve of the tangent modulus E(ε).

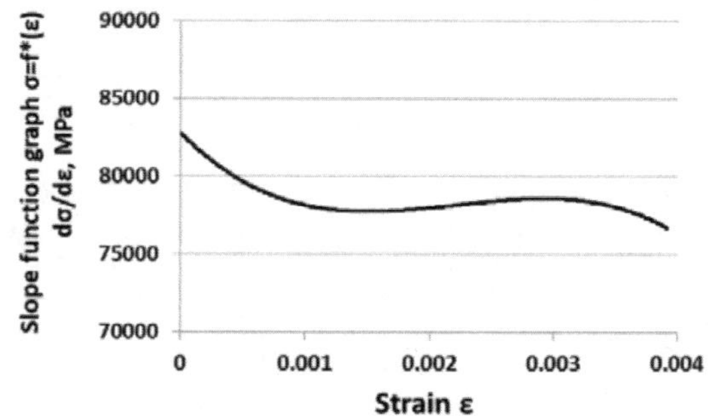

Figure 10. Conditional stress-strain diagram $\sigma = f^*(\varepsilon)$: slope tangent $d\sigma/d\varepsilon$.

The second derivative of the function (4) at the strain ε (figure 11) characterizes the direction of the convexity (↑up, ↓down) of the function graph, and also allows us to determine the inflection points (ip1, ip2), in which the direction of the convexity changes:

$$d^2\sigma/d\varepsilon^2 = -1.842\cdot 10^{12}\cdot\varepsilon^2 + 8.148\cdot 10^9\cdot\varepsilon - 8.092\cdot 10^6 \quad (6)$$

Figure 12 compares the slope of the function graph $\sigma = f^*(\varepsilon)$ to the positive direction of the strain axis equation (5) with the curve of the tangent modulus E(ε) equation (3) obtained by the methods described above. There are no qualitative differences and there is an insignificant relative shift in the strain of the second inflection points (ip2) of the graphs of the functions $d\sigma/d\varepsilon$ and E(ε). The maximum (at $\varepsilon = 0$) quantitative difference is about 2.5%, which indicates that it is acceptable to use both methods to identify the nonlinearity of the stress – strain dependence of a quasi-isotropic structural FCRP laminate.

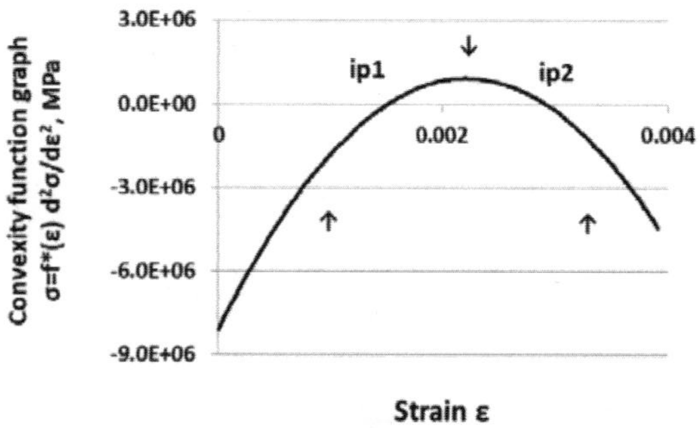

Figure 11. Conditional stress-strain diagram $\sigma=f^*(\varepsilon)$: direction of convexity and points of inflection $d^2\sigma/d\varepsilon^2$.

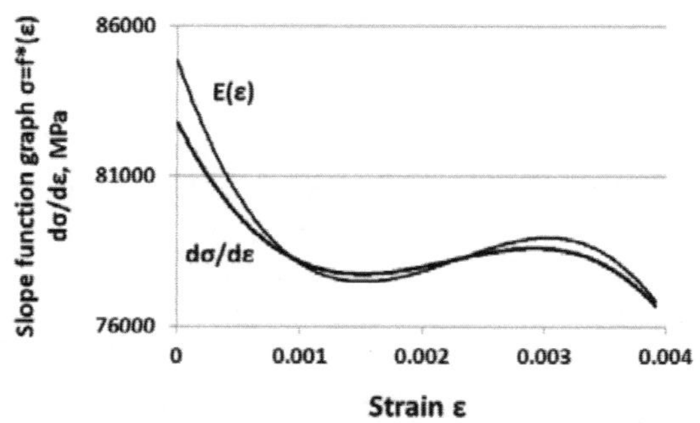

Figure 12. Conditional stress-strain diagram $\sigma= f^*(\varepsilon)$: first derivative $d\sigma/d\varepsilon$ and the tangent modulus E(ε).

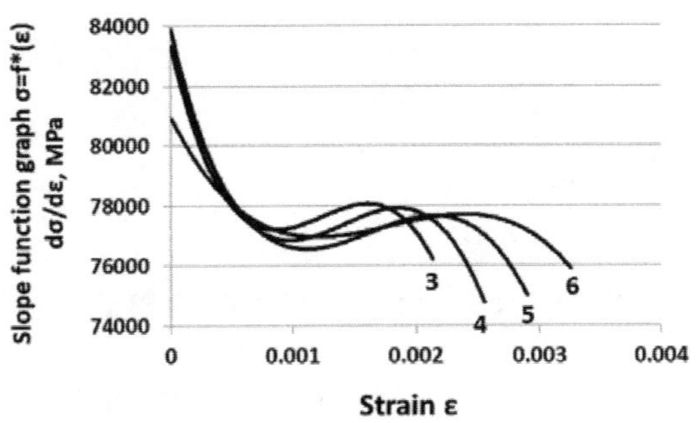

Figure 13. Conditional stress-strain diagram σ = f*(ε) in loading cycles 3–6: slope tangent dσ/dε.

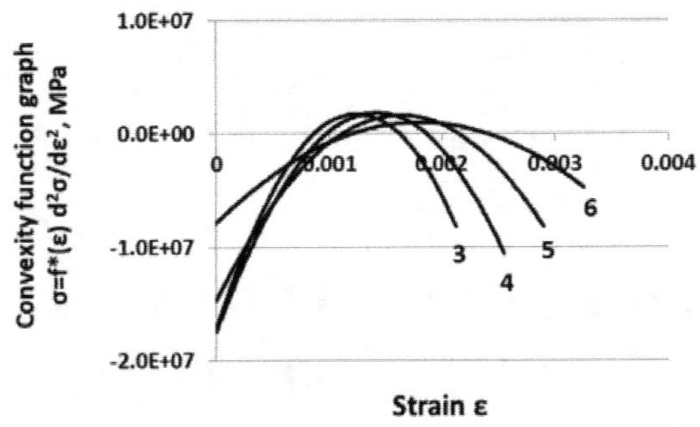

Figure 14. Conditional stress-strain diagram σ= f*(ε) in loading cycles 3–6: direction of convexity and points of inflection $d^2\sigma/d\varepsilon^2$.

4. Conclusions

The conducted studies of the experimental stress-strain dependence for a specimen of quasi-isotropic structural FCRP laminate in one of the cycles of re-static loading allowed to establish the nonlinearity of this dependence, to determine the type of curve approximating the experimental data, and to minimize the number of nonlinearity parameters of the approximating curve – the slope of the graph, the direction of the convexity and the inflection points.

In the future, it is intended to use the change of the nonlinear part of the stress – strain dependence as an indicator of the degradation of material properties under re-static loading (figures 13 and 14) [4].

References

[1] Makarov P V and Eremin M O 2013 *Physical Mesomechanics* **16(3)** pp 207-26
[2] Rabotnov Y N 1979 *Mechanics of deformable solids* (Moscow: Nauka)
[3] Feodosiev V I 1974 *Strength of materials* (Moscow: Nauka)
[4] Severov P B 2016 *Engineering and Automation Problems* **4** pp 85-92

Damaged zone thickness estimation method for machined FRP composite specimens

D D Vlasov[1], N A Tatus'[2], A N Polilov[2] and E Kh Akhmedshin[3]

[1] Bauman Moscow State Technical University, 2nd Baumanskaya str, 5, Moscow, 105005, Russia
[2] Mechanical Engineering Research Institute of the Russian Academy of Sciences, 4 Maly Kharitonyevsky Pereulok, Moscow, 101990, Russia
[3] Moscow Polytech, Bolshaya Semenovskaya str, 38, Moscow, 107023, Russia

polilovan@mail.ru

Abstract. The results of the experimental study of machining effect on the fiberglass specimens' tensile strength realization are presented. According to the data obtained, the size of the damaged zone, which does not carry applied loads and reduces the actual net cross-section of the specimen, has been estimated.

1. Introduction

Fibrous composite materials have a heterogeneous structure where strong and brittle fibers, which are easily damaged during machining, play a major bearing role. The damage may not be visible, but if for metals machining process leads to local hardening and microplastic deformations that have little effect on strength, then for glass and carbon plastics machining damage can have a significant negative effect [1]. Due to this fact machining on the thickness of composite plates is not allowed, and the technology and cutting modes of linear specimens have been carefully worked out.

The need to cut composite materials along curved trajectories required the acquisition and adjustment of a modern, imported CNC 6040Z milling machine, which allows you to direct the cutting tool along any trajectory, and not only straight.

2. Setting the task

When working out cutting modes, several questions arise: How does cutting technology affect the damage of material near the cutting line? How can we estimate the thickness of damaged layer of material that actually goes out of bearing part of specimen?

To answer these questions, at the first stage it was decided to do a tensile test of rectangular specimens with different width that have been cut out with the single technology. This made it possible to assess the size of damaged zone - the "edge," which is formally the part of the cross section of the specimen, but does not bear any load. It is considered that in a rectangular specimen cut from the plate with a width of b and a length of L the width of damaged area δ (figure 1) remains constant.

It should be noted that the width of the "edge" must be constant for a given size and type of cutting tool, as well as at fixed cutting modes (rotation and feed speed). Therefore, with an increase of the size of the structure (sample width), the relative width of the damaged edge and its effect on strength should be reduced.

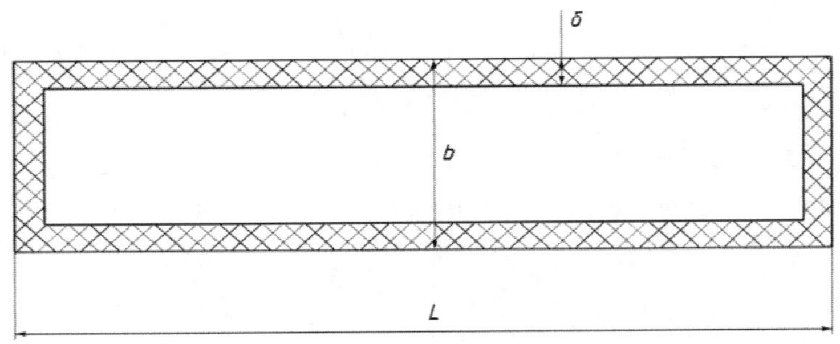

Figure 1. The damaged zone of specimen after machining process.

3. Specimen fabrication and testing

The fiberglass plate with the thickness of h = 2 mm has been made with the vacuum infusion technology. The plate has been molded from 6 layers of glass fabric with a weave type of "web" and laying [0/90].

Specimens with a length of 250 mm with different widths have been cut from a plate on the CNC milling machine using a corn cutter tool (figure 2) with a diameter of 1.5 mm. Horizontal feed rate: 250 mm·min^{-1}, inverter current frequency: 160 Hz. The cutting mode for all specimens has not changed.

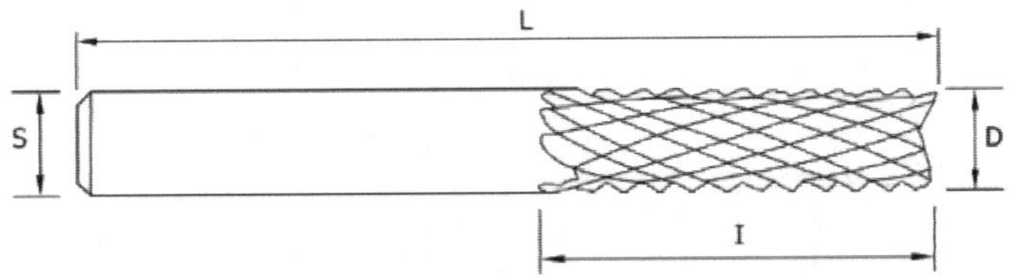

Figure 2. Corn type cutter tool.

For the test implementation 7 series of three specimens with different widths have been made: 2.5; 3; 4; 5; 7.5; 10 and 30 mm. Series number i = 1, 2,..., 7.

The specimens have been tensile tested on an INSTRON machine in self-tightening grips. All specimens have collapsed in the working area, which indicates the correctness of the test results.

4. Test results

According to the obtained experimental data, the dependence of strength of the specimen on its width was built. Figure 3 shows that when the width of the specimen increases to 10 mm, the conditional critical stresses increase sharply, and then they go to the "plateau" with a value of about 400 MPa. This behavior of the diagram shows a strong effect of the damaged zone on the conditional strength of small-width specimens and a much smaller effect for wider ones.

To estimate the width of the damaged zone, we use simple formulas for conditional and real ("true") strength.

Conditional strength (critical load to full cross section):

$$\sigma_i^{eff} = \frac{F_i}{b_i \cdot h} \tag{1}$$

Real strength from test (load to unknown net cross section):

$$\sigma_i^{real} = \frac{F_i}{b'_i \cdot h} \tag{2}$$

where $b'_i = b_i - 2\delta -$ is the effective (actually resisting applied force) width.

For approximate calculations, the maximum value of the conditional strength of this material can be used as the real ("true") strength. On the basis of the diagram (figure 3), the "true" strength has been taken as the value of the conditional strength obtained on specimens with the width of 30 mm since at this width the damaged zone no longer has a significant impact on the implementation of the strength.

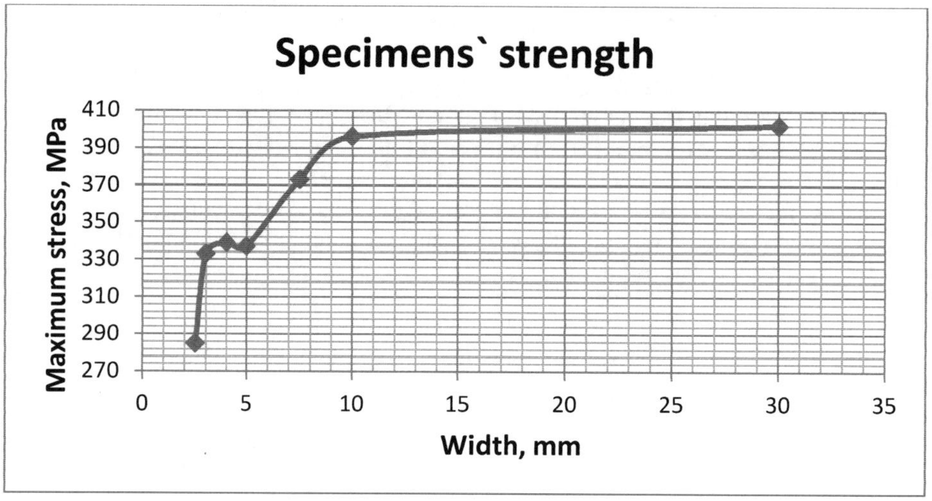

Figure 3. Conditional strength on width dependence.

Changing σ_i^{real} to $\sigma^* = \sigma_7^{eff} = 402MPa$ in (2) (index «7» means the width of specimen b_7 = 30 mm) we estimate from the experimental values of the critical load F_i the width of the damage zone δ, which turns out to be slightly different for specimens of different widths:

$$\sigma^* = \frac{F_i}{b'_i \cdot h} \tag{3}$$

$$\delta = \frac{1}{2}\left(b_i - \frac{F_i}{\sigma^* \cdot h}\right) \tag{4}$$

The results of the evaluation according to the formula (4) of the size of the "edge" for specimens of different widths (i = 1-5), taking into account the spread, are given in figure 4. It was assumed at first that the width of the damaged zone should not depend on the width of the specimens: its average value according to the data given in figure 4 is 0.24 mm. However, in reality, the calculation according to the formula (4) gives different values, and the linear dependence on the width of the samples obtained using the least squares method is built on figure 4. What is the reason for this dependence is still difficult to explain, but it is important that the width of the damaged area for a given tool and the selected cutting mode is within the range of 0.2-0.3 mm, at least not more than 0.3 mm, which can be used to practically evaluate the effect of machining.

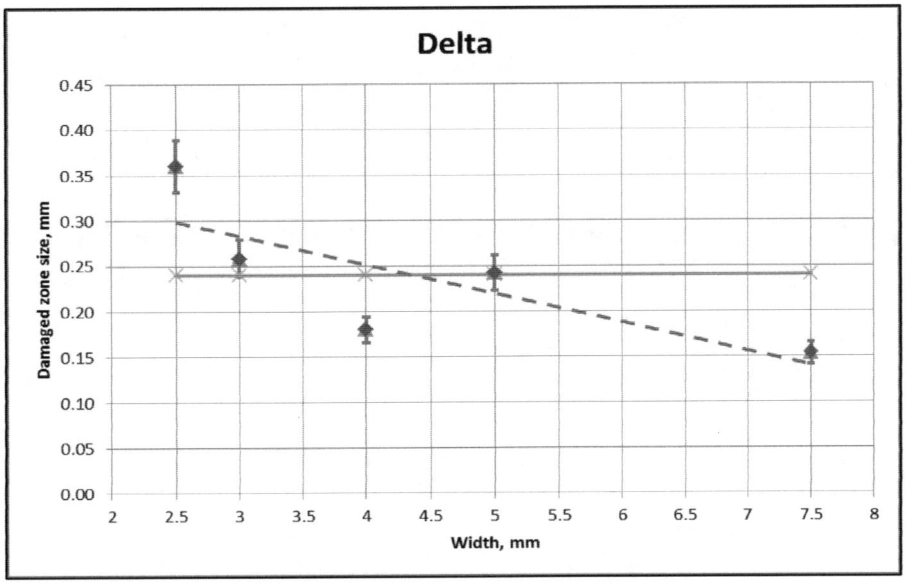

Figure 4. Damaged zone size estimation.

Results

1. The showed method can become the basis for estimating the effective width of the damaged zone when cutting composite parts. For the given data, $\delta \approx 0.24$ mm, which is about a sixth part of the diameter of the cutting tool.

2. The value of $\delta \leq 0.3$ mm corresponds only to the tool size and type used in the experiments at the selected horizontal feed rate and rotation speed. When changing the tool type and cutting modes, the effective width of the damage zone may be different.

3. The nominal size of the damaged zone is small and can be ignored for big parts, but making small parts from fibrous composites, the effect of damaged zone should be taken into account.

References

[1] Savitskiy R S and Veshkin E A 2017 Effect of mechanical processing of samples at cutting on composites testing *Izvestia of Samara Scientific Center of the Russian Academy of Sciences* vol 19 **4-2** p 214-19

Deformation and failure of polymer composites with different layouts under preliminary cyclic and quasi-static loads

V E Wildemann and O A Staroverov

[1]Perm National Research Polytechnic University, 29 Komsomolsky AVE, Perm 614990, Russia

cem_staroverov@mail.ru

Abstract. The work presents the results of an experimental study of the mechanical behavior of polymer composite materials (PCM) subjected to sequential cyclic and quasi-static loads at normal temperatures. The methodological features of testing composites under complex mechanical influences are considered. The results of tests for preliminary cyclic and subsequent quasi-static tension of fiberglass specimens with different layouts are given in the fatigue sensitivity's diagrams. Experimental data have been obtained confirming the difference in the nature of the change in the residual strength of fiberglass with different layouts.

1. Introduction

Polymer composites are increasingly finding their use as structural materials in the nodes of highly loaded parts and assemblies of the aviation, space, transport, marine and other types of industries. One of the main advantages of composites based on a polymer matrix is their lower weight with relatively equal mechanical characteristics compared to traditional structural steels and alloys. The problem of designing structures made of orthotropic materials is taking into account all possible mechanical influences applied in a direction different from the direction of laying reinforcing elements during operation. To solve this problem, it is necessary to develop new and modernize the existing scientifically grounded and standardized test methods for composite materials. The combined use of modern testing and diagnostic systems makes it possible to implement complex loading modes and obtain more data on the processes of damage accumulation and fracture kinetics. In the aggregate, the development of approaches to experimental studies of the deformation and strength properties of modern PCM under complex mechanical loads using modern electromechanical, servo-hydraulic, electrodynamic testing and diagnostic equipment for infrared thermal scanning, deformation field analysis and flaw detection is a demanded scientific and practical task. The regularities of changes in the strength and deformation properties of composite specimens with different layouts and parameters of preliminary cyclic loads are reflected in works [1-8].

2. Materials and equipment

The experimental part of the work was carried out using the Instron E10k equipment (figure 1 (a)) for preliminary cyclic exposure and Instron 8850 for quasi-static tension (figure 1 (b)). The objects of research were layered-fibrous cross-reinforced fiberglass samples with layouts of reinforcing layers $[0/90]_n$, $[0/30/0/60]_n$, $[\pm 45]_n$ based on prepreg VPS-48 and binder VSE-1212.

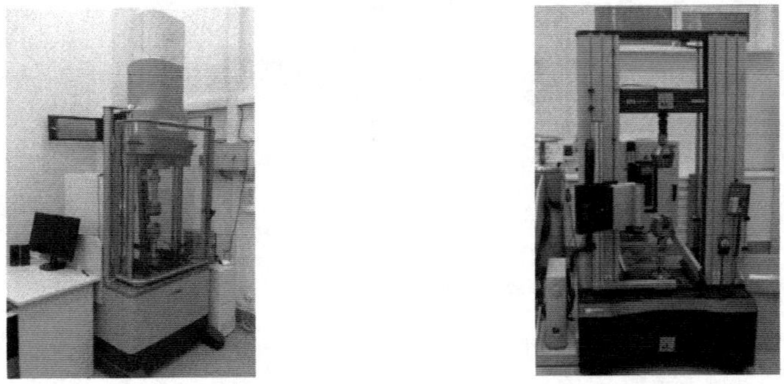

(a) (b)

Figure 1. Photo of the Instron E 10k (a) and Instron 5882 (b) test systems.

The test sequence is shown schematically in figure 2.

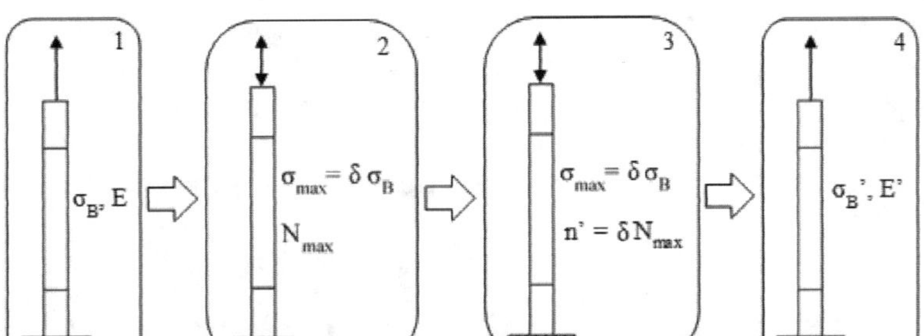

Figure 2. The research methodology for cyclic and quasi-static tension.

The test procedure includes the following stages:

1. quasi-static tests with determination of ultimate strength σ_B and elastic modulus E of the material;
2. cyclic tests at various loading parameters $\sigma_{max} = \delta \cdot \sigma_B$, where $\delta \in [0; 1]$ with the determination of the fatigue life value N_{max};
3. preliminary load with different operating time of cycles $n' = \delta \cdot N_{max}$, where $\delta \in [0; 1]$;
4. quasi-static loading with the determination of the values of the residual mechanical characteristics of the material σ_B', E'.

The shape and dimensions of the samples must comply with the recommendations of the standard ASTM D3039.

3. Results

As a result of tests for preliminary cyclic and subsequent quasi-static stretching, a difference was revealed in the nature of the change in the residual strength of fiberglass with different layouts (figure 3). The test results are presented in the form of fatigue sensitivity diagrams [9].

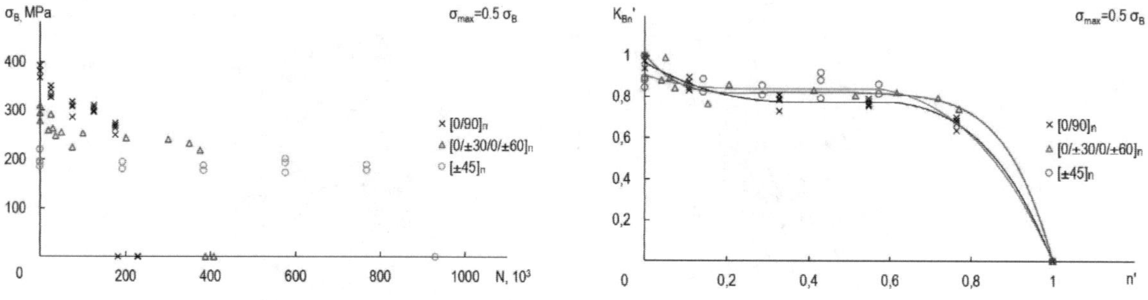

(a) (b)

Figure 3. Diagrams of the dependence of the residual strength on number of cyclic (a) and fatigue sensitivity (b) of GFRP samples with different layouts

For specimens with laying angles $[0/90]_n$ and $[0/30/0/60]_n$, three zones can be distinguished on the fatigue sensitivity diagram: initial, stabilization and sharpening of fatigue sensitivity. The change in the residual strength characteristics for specimens with the layouts $[\pm 45]_n$ occurred in two stages, without a section of initial fatigue sensitivity. The area of initial sensitivity for samples $[0/90]_n$ was (0-0.3)·n', stabilization (0.3-0.6)·n', sharpening (0.6-1)·n'; for samples $[0/30/0/60]_n$ section of initial sensitivity (0-0.2)·n', stabilization (0.2-0.7)·n', exacerbation (0.7-1)·n'.

Figure 4 shows photographs of specimens after fatigue failure. In specimens with packing $[\pm 45]_n$ in the process of cyclic loading, gradual elongation and fracture were observed in the direction of laying reinforcing fibers in accordance with the classification of standard ASTM D3039 - AGM. Samples with laying angles $[0/90]_n$ and $[0/30/0/60]_n$ fracture proceeded according to the normal tear-off scheme in accordance with the classification of the standard ASTM D3039 - LGT.

Figure 4. Photos of destroyed GFRP samples with different layouts.

4. Conclusions

A comprehensive study of the mechanical behavior of layered fiberglass samples under cyclic and sequential quasi-static tension has been carried out. New experimental data have been obtained reflecting the regularities of changes in the strength and deformation characteristics of polymer composite materials with different orientation of the reinforcement layers in the process of fatigue damage accumulation. It is shown that for fiberglass specimens with reinforcement schemes [± 45] n in comparison with specimens with schemes [0/90] n and [0/30/0/60] n, the character of change in residual strength changed from three stages to two stages.

References

[1] Brunner A J, Stelzer S, Pinter G and Terrasi G P 2016 Cyclic fatigue delamination of carbon fiber-reinforced polymer-matrix composites: Data analysis and design considerations *International J. of Fatigue* vol 83 pp 293–99 DOI:10.1016/j.ijfatigue.2015.10.025

[2] Senatov F S, Niaza K V, Stepashkin A A and Kaloshkin S D 2016 Low-cycle fatigue behavior of 3d-printed PLA-based porous scaffolds *Composites Part B: Engineering* vol 97 pp 193–

200 DOI:10.1016/j.compositesb.2016.04.067
[3] Simon I, Banks-Sills L and Fourman V 2017 Mode I delamination propagation and R-ratio effects in woven composite DCB specimens for a multi-directional layup *International J. of Fatigue* vol 96 pp 237–51 DOI:10.1016/j.ijfatigue.2016.12.005
[4] Brod M, Just G, Dean A, Jansen E, Koch I, Rolfes R and Gude M 2019 Numerical modelling and simulation of fatigue damage in carbon fibre reinforced plastics at different stress ratios *Thin-Walled Structures* vol 139 pp 219–31 DOI:10.1016/j.tws.2019.03.005
[5] Ma Y, Zhang Y, Sugahara T, Jin S, Yang Y and Hamada H 2016 Off-axis tensile fatigue assessment based on residual strength for the unidirectional 45° carbon fiber-reinforced composite at room temperature *Composites Part A: Applied Science and Manufacturing* vol 90 pp 711–23 DOI:10.1016/j.compositesa.2016.09.001
[6] Katunin A and Wachla D 2018 Determination of fatigue limit of polymeric composites in fully reversed bending loading mode using self-heating effect *J. of Composite Materials* 002199831878045 DOI:10.1177/0021998318780454
[7] Katunin A 2018 Criticality of the self-heating effect in polymers and polymer matrix composites during fatigue, and their application in non-destructive testing *Polymers* vol 11(1) pp 29 DOI:10.3390/polym11010019
[8] Post N L, Cain J, McDonald K J, Case S W and Lesko J J 2008 Residual strength prediction of composite materials: Random spectrum loading *Engineering Fracture Mechanics* vol 75 pp 2707–24
[9] Wil'deman V E, Staroverov O A and Lobanov D S 2018 Diagram and parameters of fatigue sensitivity for evaluating the residual strength of layered GFRP composites after preliminary cyclic loadings *Mechanics of Composite Materials* vol 54 **3** pp 313-20

Acknowledgements
The work was carried out in the Perm National Research Polytechnic University with support of the Russian Science Foundation (Project 16-19-00069).

AUTHOR INDEX